Introduction to Mechatronics

Introduction to Mechatronics

Edited by
Randy Dodd

Larsen & Keller
www.larsen-keller.com

Introduction to Mechatronics
Edited by Randy Dodd
ISBN: 978-1-63549-180-7 (Hardback)

© 2017 Larsen & Keller

▤ Larsen & Keller

Published by Larsen and Keller Education,
5 Penn Plaza,
19th Floor,
New York, NY 10001, USA

Cataloging-in-Publication Data

Introduction to mechatronics / edited by Randy Dodd.
    p. cm.
Includes bibliographical references and index.
ISBN 978-1-63549-180-7
1. Mechatronics. 2. Robotics. I. Dodd, Randy.
TJ163.12 .I58 2017
621--dc23

The publisher's policy is to use permanent paper from mills that operate a sustainable forestry policy. Furthermore, the publisher ensures that the text paper and cover boards used have met acceptable environmental accreditation standards.

Printed and bound in the United States of America.

For more information regarding Larsen and Keller Education and its products, please visit the publisher's website www.larsen-keller.com

Table of Contents

Preface

This book attempts to understand the multiple branches that fall under the discipline of mechatronics and how such concepts have practical applications. It talks in detail about the advancements and requirements of this field. Mechatronics is an amalgamation of various different branches of engineering like telecommunications engineering, control engineering, computer engineering, systems engineering, mechanical engineering and electronics, etc. Mechatronics is the meeting-point of various allied branches of engineering and it possesses a highly practical approach and methodology. This text is compiled in such a manner, that it will provide in-depth knowledge about the theory and practice of mechatronics. The various sub-fields along with their technological progress, that have future implications are glanced at in it. It also elucidates the concepts and innovative models around prospective developments with respect to mechatronics. As this field is emerging at a rapid pace, the contents of this textbook will help the readers understand the modern concepts and applications of the subject.

A detailed account of the significant topics covered in this book is provided below:

Chapter 1- Mechatronics is a branch that integrates mechanical engineering with electronics, computer and telecommunications engineering. It studies the design and manufacture of electronic systems and machine tools. This chapter describes the main aims of mechatronics and provides the user with a brief overview of the subject.

Chapter 2- Mechatronics utilizes several concepts that have been borrowed from the multiple disciplines that constitute it. The key concepts explained in this chapter are kinematic chain, Denavit–Hartenberg parameters, numerical control, degrees of freedom (mechanics), control theory, gimbal lock, motion planning, shortest path problem etc. The applications of these concepts in mechatronics have also been explained to the reader.

Chapter 3- This chapter is a compilation of the various allied branches of mechatronics that form an integral part of the broader subject matter. The fields described include biomechatronics, cybernetics, ecomechatronics and electromechanics. This content details the practice, technology and methodology of each of these fields.

Chapter 4- The encompassing nature and flexibility of mechatronics has ensured its usage in applications that span disciplines. This chapter describes topics like machine vision, expert system, cyber-physical system, anti-lock breaking system, servomechanism, control system, cruise control etc. It explores the methods, uses and operation of these applications.

Chapter 5- Robotics is concerned with the design, construction and application of automated machines that can replace human beings in hazardous conditions or resemble humans in appearance. This chapter introduces the reader to concepts and principles like pneumatic cylinder, mobile service systems, parallel manipulator, articulated robot, robot-assisted surgery, bio-inspired robotics and delta robot. Mechatronics is best understood in confluence with the major topics listed in the following chapter.

Chapter 6- Mechatronics and robotics together have made possible such ground-breaking inventions like Mars Exploration Rover, Curiosity (Rover), Canadarm, the da Vinci Surgical System and Sensei robotic catheter system. These mechanical marvels have revolutionized space exploration and surgery. The chapter explores these innovations and their design.

It gives me an immense pleasure to thank our entire team for their efforts. Finally in the end, I would like to thank my family and colleagues who have been a great source of inspiration and support.

Editor

Introduction to Mechatronics

Mechatronics is a branch that integrates mechanical engineering with electronics, computer and telecommunications engineering. It studies the design and manufacture of electronic systems and machine tools. This chapter describes the main aims of mechatronics and provides the user with a brief overview of the subject.

Mechatronics is a multidisciplinary field of science that includes a combination of o mechanical engineering, electronics, computer engineering, telecommunications engineering, systems engineering and control engineering. As technology advances, the subfields of engineering multiply and adapt. Mechatronics' aim is a design process that unifies these subfields. Originally, mechatronics just included the combination of mechanics and electronics, hence the word is a combination of mechanics and electronics; however, as technical systems have become more and more complex the definition has been broadened to include more technical areas.

The word "mechatronics" originated in Japanese-English and was created by Tetsuro Mori, an engineer of Yaskawa Electric Corporation. The word "mechatronics" was registered as trademark by the company in Japan with the registration number of "46-32714" in 1971. However, afterward the company released the right of using the word to public, and the word "mechatronics" spread to the rest of the world. Nowadays, the word is translated in each language and the word is considered as an essential term for industry.

French standard NF E 01-010 gives the following definition: "approach aiming at the synergistic integration of mechanics, electronics, control theory, and computer science within product design and manufacturing, in order to improve and/or optimize its functionality".

Many people treat "mechatronics" as a modern buzzword synonymous with "electromechanical engineering".

Description

A mechatronics engineer unites the principles of mechanics, electronics, and computing to generate a simpler, more economical and reliable system. The term "mechatronics" was coined by Tetsuro Mori, the senior engineer of the Japanese company Yaskawa in 1969. An industrial robot is a prime example of a mechatronics system; it includes aspects of electronics, mechanics, and computing to do its day-to-day jobs.

Engineering cybernetics deals with the question of control engineering of mechatronic systems. It is used to control or regulate such a system. Through collaboration, the mechatronic modules perform the production goals and inherit flexible and agile manufacturing properties in the production scheme. Modern production equipment consists of mechatronic modules that are integrated according to a control architecture. The most known architectures involve hierarchy, polyarchy, heterarchy, and hybrid. The methods for achieving a technical effect are described by control algorithms, which might or might not utilize formal methods in their design. Hybrid systems important to mechatronics include production systems, synergy drives, planetary exploration rovers, automotive subsystems such as anti-lock braking systems and spin-assist, and everyday equipment such as autofocus cameras, video, hard disks, and CD players.

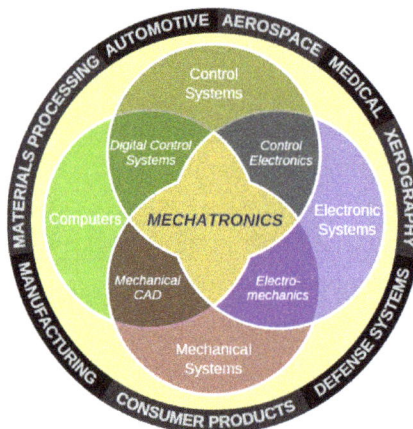

Aerial Euler diagram from RPI's website describes the s fields that make up Mechatronics

Course Structure

Mechatronic students take courses in various fields:

- Mechanical engineering and materials science
- Electrical engineering
- Computer engineering (software & hardware engineering)
- Computer science
- Systems and control engineering
- Optical engineering

Application

- Machine vision
- Automation and robotics

- Servo-mechanics

- Sensing and control systems

- Automotive engineering, automotive equipment in the design of subsystems such as anti-lock braking systems

- Computer-machine controls, such as computer driven machines like IE CNC milling machines

- Expert systems

- Industrial goods

- Consumer products

- Mechatronics systems

- Medical mechatronics, medical imaging systems

- Structural dynamic systems

- Transportation and vehicular systems

- Mechatronics as the new language of the automobile

- Computer aided and integrated manufacturing systems

- Computer-aided design

- Engineering and manufacturing systems

- Packaging

- Microcontrollers / PLCs

- Mobile apps

- M&E Engineering

Physical Implementations

Mechanical modeling calls for modeling and simulating physical complex phenomenon in the scope of a multi-scale and multi-physical approach. This implies to implement and to manage modeling and optimization methods and tools, which are integrated in a systemic approach. The specialty is aimed at students in mechanics who want to open their mind to systems engineering, and able to integrate different physics or technologies, as well as students in mechatronics who want to increase their knowledge in optimization and multidisciplinary simulation technics. The specialty educates students in robust and/or optimized conception methods for structures or many technological systems, and to the main modeling and simulation tools used in R&D. Special courses are also proposed for original applications (multi-materials composites, innovating transducers and actuators, integrated systems, …) to prepare the students to the coming breakthrough in the domains covering the materials and the systems. For some mechatronic systems, the main issue is no longer how to implement a control system, but how to implement actuators. Within the mechatronic field, mainly two technologies are used to produce movement/motion.

Variant of the Field

An emerging variant of this field is biomechatronics, whose purpose is to integrate mechanical parts with a human being, usually in the form of removable gadgets such as an exoskeleton. This is the "real-life" version of cyberware.

Another variant that we can consider is Motion control for Advanced Mechatronics, which presently is recognized as a key technology in mechatronics. The robustness of motion control will be represented as a function of stiffness and a basis for practical realization. Target of motion is parameterized by control stiffness which could be variable according to the task reference. However, the system robustness of motion always requires very high stiffness in the controller.

References

- Mechanical and Mechatronics Engineering Department. "What is Mechatronics Engineering?". Prospective Student Information. University of Waterloo. Retrieved 30 May 2011.

- Faculty of Mechatronics, Informatics and Interdisciplinary Studies TUL. "Mechatronics (Bc., Ing., PhD.)". Retrieved 15 April 2011.

- Lawrence J. Kamm (1996). Understanding Electro-Mechanical Engineering: An Introduction to Mechatronics. John Wiley & Sons. ISBN 978-0-7803-1031-5.

Key Concepts of Mechatronics

Mechatronics utilizes several concepts that have been borrowed from the multiple disciplines that constitute it. The key concepts explained in this chapter are kinematic chain, Denavit–Hartenberg parameters, numerical control, degrees of freedom (mechanics), control theory, gimbal lock, motion planning, shortest path problem etc. The applications of these concepts in mechatronics have also been explained to the reader.

Kinematic Chain

Kinematic chain refers to an assembly of rigid bodies connected by joints that is the mathematical model for a mechanical system. As in the familiar use of the word chain, the rigid bodies, or links, are constrained by their connections to other links. An example is the simple open chain formed by links connected in series, like the usual chain, which is the kinematic model for a typical robot manipulator.

Mathematical models of the connections, or joints, between two links are termed kinematic pairs. Kinematic pairs model the hinged and sliding joints fundamental to robotics, often called *lower pairs* and the surface contact joints critical to cams and gearing, called *higher pairs*. These joints are generally modeled as holonomic constraints. A kinematic diagram is a schematic of the mechanical system that shows the kinematic chain.

The modern use of kinematic chains includes compliance that arises from flexure joints in precision mechanisms, link compliance in compliant mechanisms and micro-electro-mechanical systems, and cable compliance in cable robotic and tensegrity systems.

The JPL mobile robot ATHLETE is a platform with six serial chain legs ending in wheels.

The arms, fingers and head of the JSC Robonaut are modeled as kinematic chains.

The movement of the Boulton & Watt steam engine is studied as a system of rigid bodies connected by joints forming a kinematic chain.

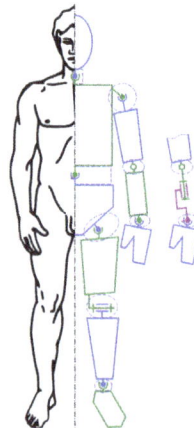

A model of the human skeleton as a kinematic chain allows positioning using forward and inverse kinematics.

Mobility Formula

The degrees of freedom, or *mobility,* of a kinematic chain is the number of parameters that define the configuration of the chain. A system of n rigid bodies moving in space has *6n* degrees of freedom measured relative to a fixed frame. This frame is included in the count of bodies, so that mobility does not depend on link that forms the fixed frame. This means the degree-of-freedom of this system is M=6(N-1), where N=n+1 is the number of moving bodies plus the fixed body.

Joints that connect bodies impose constraints. Specifically, hinges and sliders each impose five constraints and therefore remove five degrees of freedom. It is convenient to define the number of constraints *c* that a joint imposes in terms of the joint's freedom *f*, where *c=6-f*. In the case of a hinge or slider, which are one degree of freedom joints, have *f=1* and therefore *c=6-1=5*.

The result is that the mobility of a kinematic chain formed from *n* moving links and *j* joints each with freedom f_i, *i=1, ..., j*, is given by

$$M = 6n - \sum_{i=1}^{j}(6 - f_i) = 6(N - 1 - j) + \sum_{i=1}^{j}f_i$$

Recall that *N* includes the fixed link.

Analysis of Kinematic Chains

The constraint equations of a kinematic chain couple the range of movement allowed at each joint to the dimensions of the links in the chain, and form algebraic equations that are solved to determine the configuration of the chain associated with specific values of input parameters, called degrees of freedom.

The constraint equations for a kinematic chain are obtained using rigid transformations [Z] to characterize the relative movement allowed at each joint and separate rigid transformations [X] to define the dimensions of each link. In the case of a serial open chain, the result is a sequence of rigid transformations alternating joint and link transformations from the base of the chain to its end link, which is equated to the specified position for the end link. A chain of *n* links connected in series has the kinematic equations,

$$[T] = [Z_1][X_1][Z_2][X_2]...[X_{n-1}][Z_n],$$

where [T] is the transformation locating the end-link---notice that the chain includes a "zeroth" link consisting of the ground frame to which it is attached. These equations are called the forward kinematics equations of the serial chain.

Kinematic chains of a wide range of complexity are analyzed by equating the kinematics equations of serial chains that form loops within the kinematic chain. These equations are often called *loop equations*.

The complexity (in terms of calculating the forward and inverse kinematics) of the chain is determined by the following factors:

- Its topology: a serial chain, a parallel manipulator, a tree structure, or a graph.

- Its geometrical form: how are neighbouring joints spatially connected to each other?

Explanation:-

Two or more rigid bodies in space are collectively called a rigid body system. We can hinder the motion of these independent rigid bodies with kinematic constraints. Kinematic constraints are constraints between rigid bodies that result in the decrease of the degrees of freedom of rigid body system.

Synthesis of Kinematic Chains

The constraint equations of a kinematic chain can be used in reverse to determine the dimensions of the links from a specification of the desired movement of the system. This is termed *kinematic synthesis.*

Perhaps the most developed formulation of kinematic synthesis is for four-bar linkages, which is known as Burmester theory.

Ferdinand Freudenstein is often called the father of modern kinematics for his contributions to the kinematic synthesis of linkages beginning in the 1950s. His use of the newly developed computer to solve *Freudenstein's equation* became the prototype of computer-aided design systems.

This work has been generalized to the synthesis of spherical and spatial mechanisms.

Denavit–hartenberg Parameters

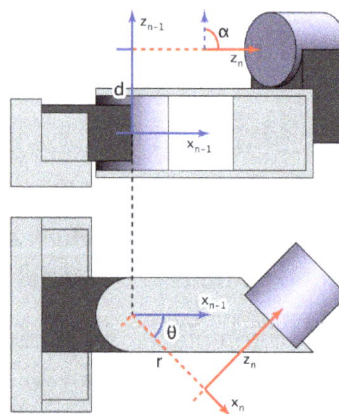

The Denavit–Hartenberg parameters (also called DH parameters) are the four parameters associated with a particular convention for attaching reference frames to the links of a spatial kinematic chain, or robot manipulator.

Jacques Denavit and Richard Hartenberg introduced this convention in 1955 in order to standardize the coordinate frames for spatial linkages.

Richard Paul demonstrated its value for the kinematic analysis of robotic systems in 1981. While many conventions for attaching references frames have been developed, the Denavit-Hartenberg convention remains a popular approach.

Denavit-hartenberg Convention

A commonly used convention for selecting frames of reference in robotics applications is the Denavit and Hartenberg (D–H) convention which was introduced by Jacques Denavit and Richard S. Hartenberg. In this convention, coordinate frames are attached to the joints between two links such that one transformation is associated with the joint, [Z], and the second is associated with the link [X]. The coordinate transformations along a serial robot consisting of n links form the kinematics equations of the robot,

$$[T] = [Z_1][X_1][Z_2][X_2]...[X_{n-1}][Z_n],$$

where [T] is the transformation locating the end-link.

In order to determine the coordinate transformations [Z] and [X], the joints connecting the links are modeled as either hinged or sliding joints, each of which have a unique line S in space that forms the joint axis and define the relative movement of the two links. A typical serial robot is characterized by a sequence of six lines S_i, $i=1,...,6$, one for each joint in the robot. For each sequence of lines S_i and S_{i+1}, there is a common normal line $A_{i,i+1}$. The system of six joint axes S_i and five common normal lines $A_{i,i+1}$ form the kinematic skeleton of the typical six degree of freedom serial robot. Denavit and Hartenberg introduced the convention that Z coordinate axes are assigned to the joint axes S_i and X coordinate axes are assigned to the common normals $A_{i,i+1}$.

This convention allows the definition of the movement of links around a common joint axis S_i by the screw displacement,

$$[Z_i] = \begin{bmatrix} \cos\theta_i & -\sin\theta_i & 0 & 0 \\ \sin\theta_i & \cos\theta_i & 0 & 0 \\ 0 & 0 & 1 & d_i \\ 0 & 0 & 0 & 1 \end{bmatrix},$$

where θ_i is the rotation around and d_i is the slide along the Z axis---either of the parameters can be constants depending on the structure of the robot. Under this convention the dimensions of each link in the serial chain are defined by the screw displacement around the common normal $A_{i,i+1}$ from the joint S_i to S_{i+1}, which is given by

$$[X_i] = \begin{bmatrix} 1 & 0 & 0 & r_{i,i+1} \\ 0 & \cos\alpha_{i,i+1} & -\sin\alpha_{i,i+1} & 0 \\ 0 & \sin\alpha_{i,i+1} & \cos\alpha_{i,i+1} & 0 \\ 0 & 0 & 0 & 1 \end{bmatrix},$$

where $\alpha_{i,i+1}$ and $r_{i,i+1}$ define the physical dimensions of the link in terms of the angle measured around and distance measured along the X axis.

In summary, the reference frames are laid out as follows:

1. z -axis is in the direction of the joint axis

2. x -axis is parallel to the common normal: $x_n = z_{n-1} \times z_n$ If there is no unique common normal (parallel z axes), d (below) is a free parameter. The direction of x_n is z_{n-1} z_n , as shown in the video below.

3. the y -axis follows from the x - and z -axis by choosing it to be a right-handed coordinate system.

Four Parameters

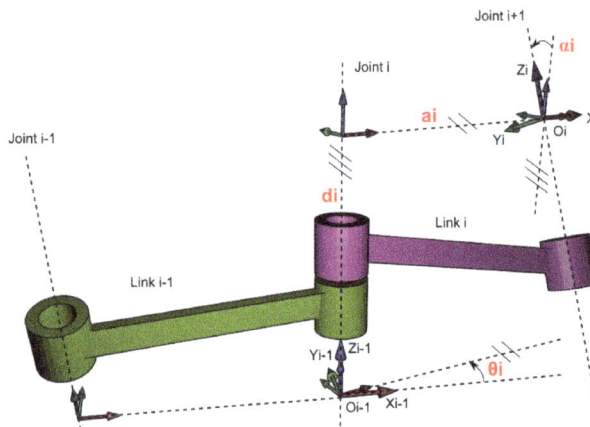

The four parameters of classic DH convention are shown in red text, which $\theta_i, d_i, a_i, \alpha_i$. With those four parameters, we can translate the coordinates from $O_{i-1}X_{i-1}Y_{i-1}Z_{i-1}$ to $O_iX_iY_iZ_i$.

The following four transformation parameters are known as D−H parameters:.

* d : offset along previous z to the common normal

* θ : angle about previous z , from old x to new x

* r : length of the common normal (aka a , but if using this notation, do not confuse with α). Assuming a revolute joint, this is the radius about previous z .

* α : angle about common normal, from old z axis to new z axis

A visualization of D–H parameterization is available: YouTube

There is some choice in frame layout as to whether the previous x axis or the next x points along the common normal. The latter system allows branching chains more efficiently, as multiple frames can all point away from their common ancestor, but in the alternative layout the ancestor can only point toward one successor. Thus the commonly used notation places each down-chain x axis collinear with the common normal, yielding the transformation calculations shown below.

We can note constraints on the relationships between the axes:

- the x_n-axis is perpendicular to both the z_{n-1} and z_n axes

- the x_n-axis intersects both z_{n-1} and z_n axes

- the origin of joint n is at the intersection of x_n z_n

- y_n completes a right-handed reference frame based on x_n and z_n

Denavit-hartenberg Matrix

It is common to separate a screw displacement into the product of a pure translation along a line and a pure rotation about the line, so that

$$[Z_i] = \text{Trans}_{Z_i}(d_i)\,\text{Rot}_{Z_i}(\theta_i),$$

and

$$[X_i] = \text{Trans}_{X_i}(r_{i,i+1})\,\text{Rot}_{X_i}(\alpha_{i,i+1}).$$

Using this notation, each link can be described by a coordinate transformation from the previous coordinate system to the next coordinate system.

$$^{n-1}T_n = \text{Trans}_{z_{n-1}}(d_n)\cdot\text{Rot}_{z_{n-1}}(\theta_n)\cdot\text{Trans}_{x_n}(r_n)\cdot\text{Rot}_{x_n}(\alpha_n)$$

Note that this is the product of two screw displacements, The matrices associated with these operations are:

$$\text{Trans}_{z_{n-1}}(d_n) = \begin{bmatrix} 1 & 0 & 0 & 0 \\ 0 & 1 & 0 & 0 \\ 0 & 0 & 1 & d_n \\ 0 & 0 & 0 & 1 \end{bmatrix}$$

$$\text{Rot}_{z_{n-1}}(\theta_n) = \begin{bmatrix} \cos\theta_n & -\sin\theta_n & 0 & 0 \\ \sin\theta_n & \cos\theta_n & 0 & 0 \\ 0 & 0 & 1 & 0 \\ 0 & 0 & 0 & 1 \end{bmatrix}$$

$$\text{Trans}_{x_n}(r_n) = \begin{bmatrix} 1 & 0 & 0 & r_n \\ 0 & 1 & 0 & 0 \\ 0 & 0 & 1 & 0 \\ 0 & 0 & 0 & 1 \end{bmatrix}$$

$$\text{Rot}_{x_n}(\alpha_n) = \begin{bmatrix} 1 & 0 & 0 & 0 \\ 0 & \cos\alpha_n & -\sin\alpha_n & 0 \\ 0 & \sin\alpha_n & \cos\alpha_n & 0 \\ 0 & 0 & 0 & 1 \end{bmatrix}$$

This gives:

$$^{n-1}T_n = \begin{bmatrix} \cos\theta_n & -\sin\theta_n\cos\alpha_n & \sin\theta_n\sin\alpha_n & r_n\cos\theta_n \\ \sin\theta_n & \cos\theta_n\cos\alpha_n & -\cos\theta_n\sin\alpha_n & r_n\sin\theta_n \\ 0 & \sin\alpha_n & \cos\alpha_n & d_n \\ 0 & 0 & 0 & 1 \end{bmatrix} = \begin{bmatrix} & R & & T \\ 0 & 0 & 0 & 1 \end{bmatrix}$$

where R is the 3×3 submatrix describing rotation and T is the 3×1 submatrix describing translation.

Use of Denavit and Hartenberg Matrices

The Denavit and Hartenberg notation gives a standard methodology to write the kinematic equations of a manipulator. This is specially useful for serial manipulators where a matrix is used to represent the pose (position and orientation) of one body with respect to another.

The position of body n with respect to $n-1$ may be represented by a position matrix indicated with the T or M

$$^{n-1}T_n = M_{n-1,n}$$

This matrix is also used to transform a point from n to $n-1$

$$M_{n-1,n} = \begin{bmatrix} R_{xx} & R_{xy} & R_{xz} & T_x \\ R_{yx} & R_{yy} & R_{yz} & T_y \\ R_{zx} & R_{zy} & R_{zz} & T_z \\ 0 & 0 & 0 & 1 \end{bmatrix}$$

Where the upper left 3×3 submatrix of M represents the relative orientation of the two bodies, and the upper 3×1 represents their relative position.

The position of k with respect to i can be obtained as the product of the matrices representing the pose of j with respect of i and that k with respect of j

$$M_{i,k} = M_{i,j} M_{j,k}$$

An important property of Denavit and Hartenberg matrices is that the inverse is

$$M^{-1} = \left[\begin{array}{ccc|c} & R^T & & -R^T T \\ \hline 0 & 0 & 0 & 1 \end{array} \right]$$

where R^T is both the transpose and the inverse of the orthogonal matrix R, i.e

$$R_{ij}^{-1} = R_{ij}^T = R_{ji}.$$

Kinematics

Further matrices can be defined to represent velocity and acceleration of bodies. The velocity of i with respect to j can be represented in frame k by the matrix

$$W_{i,j(k)} = \left[\begin{array}{ccc|c} 0 & -\omega_z & \omega_y & v_x \\ \omega_z & 0 & -\omega_x & v_y \\ -\omega_y & \omega_x & 0 & v_z \\ \hline 0 & 0 & 0 & 0 \end{array} \right]$$

where ω is the angular velocity of j with respect to i and all the components are expressed in frame k; v is the velocity of one point of body j with respect to i (the pole). The pole is the point of j passing through the origin of i.

The acceleration matrix can be defined as the sum of the time derivative of the velocity plus the velocity squared

$$H_{i,j(k)} = \dot{W}_{i,j(k)} + W_{i,j(k)}^2$$

The velocity and the acceleration in frame i of a point of body j can be evaluated as

$$\dot{P} = W_{i,j} P$$

$$\ddot{P} = H_{i,j} P$$

It is also possible to prove that

$$\dot{M}_{i,j} = W_{i,j(i)} M_{i,j}$$

$$\ddot{M}_{i,j} = H_{i,j(i)} M_{i,j}$$

Velocity and acceleration matrices add up according to the following rules

$$W_{i,k} = W_{i,j} + W_{j,k}$$

$$H_{i,k} = H_{i,j} + H_{j,k} + 2W_{i,j} W_{j,k}$$

in other words the absolute velocity is the sum of the drag plus the relative velocity; for the acceleration the Coriolis' term is also present.

The components of velocity and acceleration matrices are expressed in an arbitrary frame k and transform from one frame to another by the following rule

$$W_{(h)} = M_{h,k} W_{(k)} M_{k,h}$$

$$H_{(h)} = M_{h,k} H_{(k)} M_{k,h}$$

Dynamics

For the dynamics 3 further matrices are necessary to describe the inertia J, the linear and angular momentum Φ, and the forces and Γ applied to a body.

Inertia J:

$$J = \begin{bmatrix} I_{xx} & I_{xy} & I_{xz} & x_g m \\ I_{yx} & I_{yy} & I_{yz} & y_g m \\ I_{xz} & I_{yz} & I_{zz} & z_g m \\ x_g m & y_g m & z_g m & m \end{bmatrix}$$

where m is the mass, x_g, y_g, z_g represent the position of the center of mass, and the terms $I_{xx}, I_{xy}, \ldots$ represent inertia and are defined as

$$I_{xx} = \int \int x^2 \, dm$$

$$I_{xy} = \int \int xy \, dm$$

$$I_{xz} = \cdots$$
$$\cdots$$

Action matrix Φ, containing f and torque t:

$$\Phi = \begin{bmatrix} 0 & -t_z & t_y & f_x \\ t_z & 0 & -t_x & f_y \\ -t_y & t_x & 0 & f_z \\ -f_x & -f_y & -f_z & 0 \end{bmatrix}$$

Momentum matrix Γ, containing ρ and angular γ momentum

$$\Gamma = \left[\begin{array}{ccc|c} 0 & -\gamma_z & \gamma_y & \rho_x \\ \gamma_z & 0 & -\gamma_x & \rho_y \\ -\gamma_y & \gamma_x & 0 & \rho_z \\ \hline -\rho_x & -\rho_y & -\rho_z & 0 \end{array}\right]$$

All the matrices are represented with the vector components in a certain frame k. Transformation of the components from frame k to frame h follows to rule

$$J_{(h)} = M_{h,k} J_{(k)} M_{h,k}^{T}$$

$$\Gamma_{(h)} = M_{h,k} \Gamma_{(k)} M_{h,k}^{T}$$

$$\Phi_{(h)} = M_{h,k} \Phi_{(k)} M_{h,k}^{T}$$

The matrices described allow the writing of the dynamic equations in a concise way.

Newton's law:

$$\Phi = HJ - JH^{t}$$

Momentum:

$$\Gamma = WJ - JW^{t}$$

The first of these equations express the Newton's law and is the equivalent of the vector equation $f = ma$ (force equal mass times acceleration) plus $t = J\dot{\omega} + \omega \times J\omega$ (angular acceleration in function of inertia and angular velocity); the second equation permits the evaluation of the linear and angular momentum when velocity and inertia are known.

Modified DH Parameters

Some books such as use modified DH parameters. The difference between the classic DH parameters and the modified DH parameters are the locations of the coordinates system attachment to the links and the order of the performed transformations.

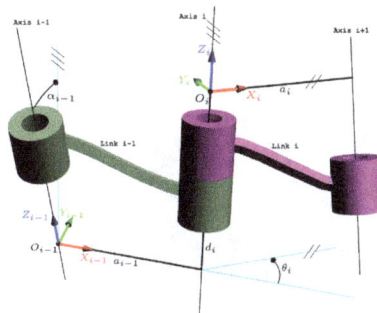

Modified DH Parameters

Compared with the classic DH parameters, the coordinates of frame O_{i-1} is put on axis i-1, not the axis i in classic DH convention. The coordinates of O_i is put on the axis i, not the axis i+1 in classic DH convention.

Another difference is that according to the modified convention, the transform matrix is given by the following order of operations:

$$^{n-1}T_n = \text{Rot}_{x_{n-1}}(\alpha_{n-1}) \cdot \text{Trans}_{x_{n-1}}(a_{n-1}) \cdot \text{Rot}_{z_n}(\theta_n) \cdot \text{Trans}_{z_n}(d_n)$$

Thus, the matrix of the modified DH parameters becomes

$$^{n-1}T_n = \begin{bmatrix} \cos\theta_n & -\sin\theta_n & 0 & a_{n-1} \\ \sin\theta_n \cos\alpha_{n-1} & \cos\theta_n \cos\alpha_{n-1} & -\sin\alpha_{n-1} & -d_n \sin\alpha_{n-1} \\ \sin\theta_n \sin\alpha_{n-1} & \cos\theta_n \sin\alpha_{n-1} & \cos\alpha_{n-1} & d_n \cos\alpha_{n-1} \\ 0 & 0 & 0 & 1 \end{bmatrix}$$

It should be noteworthy to remark that some books (e.g.) use a_n and α_n to indicate the length and twist of link n-1 rather than link n. As a consequence, $^{n-1}T_n$ is formed only with parameters using the same subscript.

Surveys of DH conventions and its differences have been published.

Numerical Control

Computer Numeric Control (CNC) is the automation of machine tools that are operated by precisely programmed commands encoded on a storage medium (computer command module, usually located on the device) as opposed to controlled manually by hand wheels or levers, or mechanically automated by cams alone. Most NC today is computer (or computerized) numerical control (CNC), in which computers play an integral part of the control.

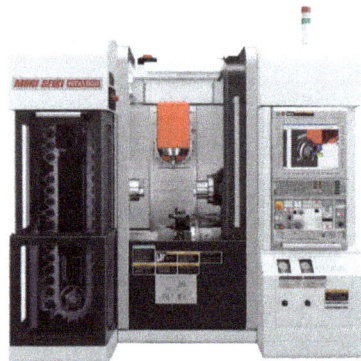

A CNC turning center

In modern CNC systems, end-to-end component design is highly automated using computer-aided design (CAD) and computer-aided manufacturing (CAM) programs. The programs produce a computer file that is interpreted to extract the commands needed to operate a particular machine by use of a post processor, and then loaded into the CNC machines for production. Since any particular component might require the use of a number of different tools – drills, saws, etc. – modern machines often combine multiple tools into a single "cell". In other installations, a number of different machines are used with an external controller and human or robotic operators that move the component from machine to machine. In either case, the series of steps needed to produce any part is highly automated and produces a part that closely matches the original CAD design.

History

The first NC machines were built in the 1940s and 1950s, based on existing tools that were modified with motors that moved the controls to follow points fed into the system on punched tape. These early servomechanisms were rapidly augmented with analog and digital computers, creating the modern CNC machine tools that have revolutionized the machining processes.

Description

Motion is controlled along multiple axes, normally at least two (X and Y), and a tool spindle that moves in the Z (depth). The position of the tool is driven by direct-drive stepper motor or servo motors in order to provide highly accurate movements, or in older designs, motors through a series of step down gears. Open-loop control works as long as the forces are kept small enough and speeds are not too great. On commercial metalworking machines, closed loop controls are standard and required in order to provide the accuracy, speed, and repeatability demanded.

As the controller hardware evolved, the mills themselves also evolved. One change has been to enclose the entire mechanism in a large box as a safety measure, often with additional safety interlocks to ensure the operator is far enough from the working piece for safe operation. Most new CNC systems built today are 100% electronically controlled.

CNC-like systems are now used for any process that can be described as a series of movements and operations. These include laser cutting, welding, friction stir welding, ultrasonic welding, flame and plasma cutting, bending, spinning, hole-punching, pinning, gluing, fabric cutting, sewing, tape and fiber placement, routing, picking and placing, and sawing.

Examples of CNC Machines

Mills

CNC mills use computer controls to cut different materials. They are able to translate programs consisting of specific numbers and letters to move the spindle (or workpiece) to various locations and depths. Many use G-code, which is a standardized programming language that many CNC machines understand, while others use proprietary languages created by their manufacturers. These proprietary languages, while often simpler than G-code, are not transferable to other machines. CNC mills have many functions including face milling, shoulder milling, tapping, drilling and some even offer turning. Standard linear CNC mills are limited to 3 axis (X, Y, and Z), but others may also have one or more rotational axes. Today, CNC mills can have 4 to 6 axes.

Lathes

A Tsugami multifunction turn mill machine used for short runs of complex parts.

Lathes are machines that cut workpieces while they are rotated. CNC lathes are able to make fast, precision cuts, generally using indexable tools and drills. They are particularly effective for complicated programs designed to make parts that would be infeasible to make on manual lathes. CNC lathes have similar control specifications to CNC mills and can often read G-code as well as the manufacturer's proprietary programming language. CNC lathes generally have two axes (X and Z), but newer models have more axes, allowing for more advanced jobs to be machined.

Plasma Cutters

CNC plasma cutting

Plasma cutting involves cutting a material using a plasma torch. It is commonly used to cut steel and other metals, but can be used on a variety of materials. In this process, gas (such as compressed air) is blown at high speed out of a nozzle; at the same time an electrical arc is formed through that gas from the nozzle to the surface being cut, turning some of that gas to plasma. The plasma is sufficiently hot to melt the material being cut and moves sufficiently fast to blow molten metal away from the cut.

Electric Discharge Machining

Electric discharge machining (EDM), sometimes colloquially also referred to as spark machining, spark eroding, burning, die sinking, or wire erosion, is a manufacturing process in which a desired shape is obtained using electrical discharges (sparks). Material is removed from the workpiece by a series of rapidly recurring current discharges between two electrodes, separated by a dielectric fluid and subject to an electric voltage. One of the electrodes is called the tool electrode, or simply the "tool" or "electrode," while the other is called the workpiece electrode, or "workpiece."

When the distance between the two electrodes is reduced, the intensity of the electric field in the space between the electrodes becomes greater than the strength of the dielectric (at the nearest point(s)), which electrically break down, allowing current to flow between the two electrodes. This phenomenon is the same as the breakdown of a capacitor. As a result, material is removed from both the electrodes. Once the current flow stops (or it is stopped – depending on the type of generator), new liquid dielectric is usually conveyed into the inter-electrode volume, enabling the solid particles (debris) to be carried away and the insulating properties of the dielectric to be restored. Adding new liquid dielectric in the inter-electrode volume is commonly referred to as flushing. Also, after a current flow, a difference of potential between the two electrodes is restored to what it was before the breakdown, so that a new liquid dielectric breakdown can occur.

Wire EDM

Also known as wire cutting EDM, wire burning EDM, or traveling wire EDM, this process uses spark erosion to machine or remove material with a traveling wire electrode from any electrically conductive material. The wire electrode usually consists of brass or zinc-coated brass material.

Sinker EDM

Sinker EDM, also called cavity type EDM or volume EDM, consists of an electrode and workpiece submerged in an insulating liquid—often oil but sometimes other dielectric fluids. The electrode and workpiece are connected to a suitable power supply, which generates an electrical potential between the two parts. As the electrode approaches the workpiece, dielectric breakdown occurs in the fluid forming a plasma channel) and a small spark jumps.

Water Jet Cutters

A water jet cutter, also known as a waterjet, is a tool capable of slicing into metal or other materials (such as granite) by using a jet of water at high velocity and pressure, or a mixture of water and an abrasive substance, such as sand. It is often used during fabrication or manufacture of parts for machinery and other devices. Waterjet is the preferred method when the materials being cut are sensitive to the high temperatures generated by other methods. It has found applications in a diverse number of industries from mining to aerospace where it is used for operations such as cutting, shaping, carving, and reaming.

Other CNC Tools

Many other tools have CNC variants, including:

- Drills
- EDMs
- Embroidery machines
- Lathes
- Milling machines
- Canned cycle
- Wood routers
- Sheet metal works (Turret punch)
- Wire bending machines
- Hot-wire foam cutters
- Plasma cutters
- Water jet cutters
- Laser cutting
- Oxy-fuel
- Surface grinders
- Cylindrical grinders

- 3D Printing

- Induction hardening machines

- Submerged welding

- Knife cutting

- Glass cutting

Tool / Machine Crashing

In CNC, a "crash" occurs when the machine moves in such a way that is harmful to the machine, tools, or parts being machined, sometimes resulting in bending or breakage of cutting tools, accessory clamps, vises, and fixtures, or causing damage to the machine itself by bending guide rails, breaking drive screws, or causing structural components to crack or deform under strain. A mild crash may not damage the machine or tools, but may damage the part being machined so that it must be scrapped.

Many CNC tools have no inherent sense of the absolute position of the table or tools when turned on. They must be manually "homed" or "zeroed" to have any reference to work from, and these limits are just for figuring out the location of the part to work with it, and aren't really any sort of hard motion limit on the mechanism. It is often possible to drive the machine outside the physical bounds of its drive mechanism, resulting in a collision with itself or damage to the drive mechanism. Many machines implement control parameters limiting axis motion past a certain limit in addition to physical limit switches. However, these parameters can often be changed by the operator.

Many CNC tools also don't know anything about their working environment. Machines may have load sensing systems on spindle and axis drives, but some do not. They blindly follow the machining code provided and it is up to an operator to detect if a crash is either occurring or about to occur, and for the operator to manually abort the cutting process. Machines equipped with load sensors can stop axis or spindle movement in response to an overload condition, but this does not prevent a crash from occurring. It may only limit the damage resulting from the crash. Some crashes may not ever overload any axis or spindle drives.

If the drive system is weaker than the machine structural integrity, then the drive system simply pushes against the obstruction and the drive motors "slip in place". The machine tool may not detect the collision or the slipping, so for example the tool should now be at 210 mm on the X axis, but is, in fact, at 32mm where it hit the obstruction and kept slipping. All of the next tool motions will be off by −178mm on the X axis, and all future motions are now invalid, which may result in further collisions with clamps, vises, or the machine itself. This is common in open loop stepper systems, but is not possible in closed loop systems unless mechanical slippage between the motor and drive mechanism has occurred. Instead, in a closed loop system, the machine will continue to attempt to move against the load until either the drive motor goes into an overcurrent condition or a servo following error alarm is generated.

Collision detection and avoidance is possible, through the use of absolute position sensors (optical encoder strips or disks) to verify that motion occurred, or torque sensors or power-draw sensors on the drive system to detect abnormal strain when the machine should just be moving and not cutting, but these are not a common component of most hobby CNC tools.

Instead, most hobby CNC tools simply rely on the assumed accuracy of stepper motors that rotate a specific number of degrees in response to magnetic field changes. It is often assumed the stepper is perfectly accurate and never missteps, so tool position monitoring simply involves counting the number of pulses sent to the stepper over time. An alternate means of stepper position monitoring is usually not available, so crash or slip detection is not possible.

Commercial CNC metalworking machines use closed loop feedback controls for axis movement. In a closed loop system, the control is aware of the actual position of the axis at all times. With proper control programming, this will reduce the possibility of a crash, but it is still up to the operator and programmer to ensure that the machine is operated in a safe manner. However, during the 2000s and 2010s, the software for machining simulation has been maturing rapidly, and it is no longer uncommon for the entire machine tool envelope (including all axes, spindles, chucks, turrets, tool holders, tailstocks, fixtures, clamps, and stock) to be modeled accurately with 3D solid models, which allows the simulation software to predict fairly accurately whether a cycle will involve a crash. Although such simulation is not new, its accuracy and market penetration are changing considerably because of computing advancements.

Numerical Precision and Equipment Backlash

Within the numerical systems of CNC programming it is possible for the code generator to assume that the controlled mechanism is always perfectly accurate, or that precision tolerances are identical for all cutting or movement directions. This is not always a true condition of CNC tools. CNC tools with a large amount of mechanical backlash can still be highly precise if the drive or cutting mechanism is only driven so as to apply cutting force from one direction, and all driving systems are pressed tightly together in that one cutting direction. However a CNC device with high backlash and a dull cutting tool can lead to cutter chatter and possible workpiece gouging. Backlash also affects precision of some operations involving axis movement reversals during cutting, such as the milling of a circle, where axis motion is sinusoidal. However, this can be compensated for if the amount of backlash is precisely known by linear encoders or manual measurement.

The high backlash mechanism itself is not necessarily relied on to be repeatedly precise for the cutting process, but some other reference object or precision surface may be used to zero the mechanism, by tightly applying pressure against the reference and setting that as the zero reference for all following CNC-encoded motions. This is similar to the manual machine tool method of clamping a micrometer onto a reference beam and adjusting the Vernier dial to zero using that object as the reference.

Degrees of Freedom (Mechanics)

In physics, the degree of freedom (DOF) of a mechanical system is the number of independent parameters that define its configuration. It is the number of parameters that determine the state of a physical system and is important to the analysis of systems of bodies in mechanical engineering, aeronautical engineering, robotics, and structural engineering.

The position of a single car (engine) moving along a track has one degree of freedom because the position of the car is defined by the distance along the track. A train of rigid cars connected by hinges to an engine still has only one degree of freedom because the positions of the cars behind the engine are constrained by the shape of the track.

An automobile with highly stiff suspension can be considered to be a rigid body traveling on a plane (a flat, two-dimensional space). This body has three independent degrees of freedom consisting of two components of translation and one angle of rotation. Skidding or drifting is a good example of an automobile's three independent degrees of freedom.

The position and orientation of a rigid body in space is defined by three components of translation and three components of rotation, which means that it has six degrees of freedom.

The exact constraint mechanical design method manages the degrees of freedom to neither underconstrain nor overconstrain a device.

Motions and Dimensions

The position of an n-dimensional rigid body is defined by the rigid transformation, $[T] = [A, d]$, where d is an n-dimensional translation and A is an $n \times n$ rotation matrix, which has n translational degrees of freedom and $n(n - 1)/2$ rotational degrees of freedom. The number of rotational degrees of freedom comes from the dimension of the rotation group SO(n).

A non-rigid or deformable body may be thought of as a collection of many minute particles (infinite number of DOFs), this is often approximated by a finite DOF system. When motion involving large displacements is the main objective of study (e.g. for analyzing the motion of satellites), a deformable body may be approximated as a rigid body (or even a particle) in order to simplify the analysis.

The degree of freedom of a system can be viewed as the minimum number of coordinates required to specify a configuration. Applying this definition, we have:

1. For a single particle in a plane two coordinates define its location so it has two degrees of freedom;

2. A single particle in space requires three coordinates so it has three degrees of freedom;

3. Two particles in space have a combined six degrees of freedom;

4. If two particles in space are constrained to maintain a constant distance from each other, such as in the case of a diatomic molecule, then the six coordinates must satisfy a single constraint equation defined by the distance formula. This reduces the degree of freedom of the system to five, because the distance formula can be used to solve for the remaining coordinate once the other five are specified.

Six Degrees of Freedom

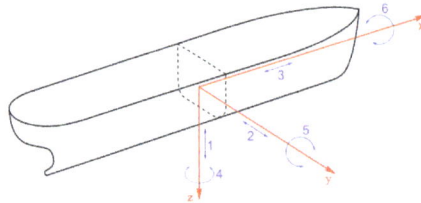

The six degrees of freedom of movement of a ship

Attitude degrees of freedom for an airplane

The motion of a ship at sea has the six degrees of freedom of a rigid body, and is described as:

Translation and Rotation:

1. Moving up and down (elevating/heaving);

2. Moving left and right (strafing/swaying);

3. Moving forward and backward (walking/surging);

4. Swivels left and right (yawing) ;

5. Tilts forward and backward (pitching);

6. Pivots side to side (rolling).

The trajectory of an airplane in flight has three degrees of freedom and its attitude along the trajectory has three degrees of freedom, for a total of six degrees of freedom.

Mobility Formula

The mobility formula counts the number of parameters that define the configuration of a set of rigid bodies that are constrained by joints connecting these bodies.

Consider a system of n rigid bodies moving in space has $6n$ degrees of freedom measured relative to a fixed frame. In order to count the degrees of freedom of this system, include the ground frame in the count of bodies, so that mobility is independent of the choice of the body that forms the fixed frame. Then the degree-of-freedom of the unconstrained system of $N = n + 1$ is

$$M = 6n = 6(N-1),$$

because the fixed body has zero degrees of freedom relative to itself.

Joints that connect bodies in this system remove degrees of freedom and reduce mobility. Specifically, hinges and sliders each impose five constraints and therefore remove five degrees of freedom. It is convenient to define the number of constraints c that a joint imposes in terms of the joint's freedom f, where $c = 6 - f$. In the case of a hinge or slider, which are one degree of freedom joints, have $f = 1$ and therefore $c = 6 - 1 = 5$.

The result is that the mobility of a system formed from n moving links and j joints each with freedom f_i, $i = 1, ..., j$, is given by

$$M = 6n - \sum_{i=1}^{j} (6 - f_i) = 6(N-1-j) + \sum_{i=1}^{j} f_i$$

Recall that N includes the fixed link.

There are two important special cases: (i) a simple open chain, and (ii) a simple closed chain. A single open chain consists of n moving links connected end to end by n joints, with one end connected to a ground link. Thus, in this case $N = j + 1$ and the mobility of the chain is

$$M = \sum_{i=1}^{j} f_i$$

For a simple closed chain, n moving links are connected end-to-end by $n + 1$ joints such that the two ends are connected to the ground link forming a loop. In this case, we have $N = j$ and the mobility of the chain is

$$M = \sum_{i=1}^{j} f_i - 6$$

An example of a simple open chain is a serial robot manipulator. These robotic systems are constructed from a series of links connected by six one degree-of-freedom revolute or prismatic joints, so the system has six degrees of freedom.

An example of a simple closed chain is the RSSR spatial four-bar linkage. The sum of the freedom of these joints is eight, so the mobility of the linkage is two, where one of the degrees of freedom is the rotation of the coupler around the line joining the two S joints.

Planar and Spherical Movement

It is common practice to design the linkage system so that the movement of all of the bodies are constrained to lie on parallel planes, to form what is known as a *planar linkage*. It is also possible to construct the linkage system so that all of the bodies move on concentric spheres, forming a *spherical linkage*. In both cases, the degrees of freedom of the links in each system is now three rather than six, and the constraints imposed by joints are now $c = 3 - f$.

In this case, the mobility formula is given by

$$M = 3(N - 1 - j) + \sum_{i=1}^{j} f_i,$$

and the special cases become

- planar or spherical simple open chain,

$$M = \sum_{i=1}^{j} f_i,$$

- planar or spherical simple closed chain,

$$M = \sum_{i=1}^{j} f_i - 3.$$

An example of a planar simple closed chain is the planar four-bar linkage, which is a four-bar loop with four one degree-of-freedom joints and therefore has mobility $M = 1$.

Systems of Bodies

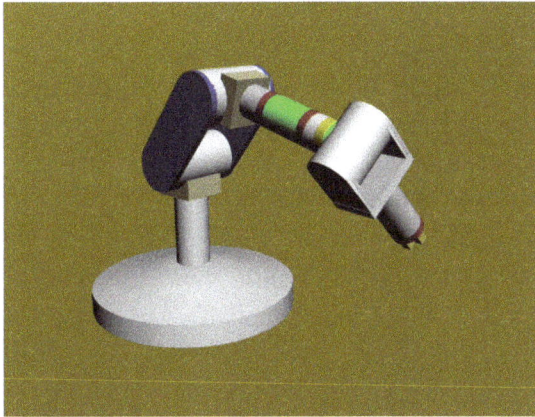

An articulated robot with six DOF in a kinematic chain.

A system with several bodies would have a combined DOF that is the sum of the DOFs of the bodies, less the internal constraints they may have on relative motion. A mechanism or linkage containing a number of connected rigid bodies may have more than the degrees of freedom for a single rigid body. Here the term *degrees of freedom* is used to describe the number of parameters needed to specify the spatial pose of a linkage.

A specific type of linkage is the open kinematic chain, where a set of rigid links are connected at joints; a joint may provide one DOF (hinge/sliding), or two (cylindrical). Such chains occur commonly in robotics, biomechanics, and for satellites and other space structures. A human arm is considered to have seven DOFs. A shoulder gives pitch, yaw, and roll, an elbow allows for pitch, and a wrist allows for pitch,yaw and roll . Only 3 of those movements would be necessary to move the hand to any point in space, but people would lack the ability to grasp things from different angles or directions. A robot (or object) that has mechanisms to control all 6 physical DOF is said to be holonomic. An object with fewer controllable DOFs than total DOFs is said to be non-holonomic, and an object with more controllable DOFs than total DOFs (such as the human arm) is said to be redundant. Although keep in mind that it is not redundant in the human arm because the two DOFs; wrist and shoulder, that represent the same movement; roll, supply each other since they can't do a full 360. The degree of freedom are like different movements that can be made.

In mobile robotics, a car-like robot can reach any position and orientation in 2-D space, so it needs 3 DOFs to describe its pose, but at any point, you can move it only by a forward motion and a steering angle. So it has two control DOFs and three representational DOFs; i.e. it is non-holonomic. A fixed-wing aircraft, with 3–4 control DOFs (forward motion, roll, pitch, and to a limited extent, yaw) in a 3-D space, is also non-holonomic, as it cannot move directly up/down or left/right.

A summary of formulas and methods for computing the degrees-of-freedom in mechanical systems has been given by Pennestri, Cavacece, and Vita.

Electrical Engineering

In electrical engineering *degrees of freedom* is often used to describe the number of directions in which a phased array antenna can form either beams or nulls. It is equal to one less than the number of elements contained in the array, as one element is used as a reference against which either constructive or destructive interference may be applied using each of the remaining antenna elements. Radar practice and communication link practice, with beam steering being more prevalent for radar applications and null steering being more prevalent for interference suppression in communication links.

Six Degrees of Freedom

Six degrees of freedom (6DoF) refers to the freedom of movement of a rigid body in three-dimensional space. Specifically, the body is free to change position as forward/backward (surge), up/down (heave), left/right (sway) translation in three perpendicular axes, combined with changes in orientation through rotation about three perpendicular axes, often termed pitch, yaw, and roll.

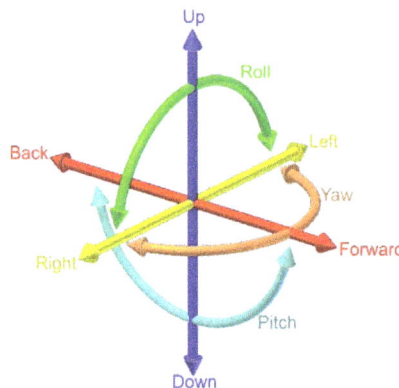

The six degrees of freedom: forward/back, up/down, left/right, pitch, yaw, roll

Robotics

Serial and parallel manipulator systems are generally designed to position an end-effector with six degrees of freedom, consisting of three in translation and three in orientation. This provides a direct relationship between actuator positions and the configuration of the manipulator defined by its forward and inverse kinematics.

Robot arms are described by their degrees of freedom. This number typically refers to the number of single-axis rotational joints in the arm, where higher number indicates an increased flexibility in positioning a tool. This is a practical metric, in contrast to the abstract definition of degrees of freedom which measures the aggregate positioning capability of a system.

In 2007, Dean Kamen, inventor of the Segway, unveiled a prototype robotic arm with 14 degrees of freedom for DARPA. Humanoid robots typically have 30 or more degrees of freedom, with six degrees of freedom per arm, five or six in each leg, and several more in torso and neck.

Engineering

The term is important in mechanical systems, especially biomechanical systems for analyzing and measuring properties of these types of systems that need to account for all six degrees of freedom. Measurement of the six degrees of freedom is accomplished today through both AC and DC magnetic or electromagnetic fields in sensors that transmit positional and angular data to a processing unit. The data are made relevant through software that integrate the data based on the needs and programming of the users.

Ascension Technology Corporation has recently created a 6DoF device small enough to fit in a biopsy needle, allowing physicians to better research at minute levels. The new sensor passively senses pulsed DC magnetic fields generated by either a cubic transmitter or a flat transmitter and is available for integration and manufacturability by medical OEMs.

An example of six degree of freedom movement is the motion of a ship at sea. It is described as :

Translation:

1. Moving forward and backward on the X-axis. (Surging)

2. Moving left and right on the Y-axis. (Swaying)

3. Moving up and down on the Z-axis. (Heaving)

Rotation

1. Tilting side to side on the X-axis. (Rolling)

2. Tilting forward and backward on the Y-axis. (Pitching)

3. Turning left and right on the Z-axis. (Yawing)

Operational Envelope Types

There are three types of operational envelope in the Six degrees of freedom. These types are *Direct*, *Semi-direct* (conditional) and *Non-direct,* all regardless of the time remaining for the execution of the maneuver, the energy remaining to execute the maneuver and finally, if the motion is commanded via a biological entity (human) or a robotical entity (computer).

1 *Direct type* : Involved a degree can be commanded directly without particularly conditions and described as a normal operation. (An aileron on a basic airplane)

2 *Semi-direct* type : Involved a degree can be commanded when some specific conditions are met. (Reverse thrust on an aircraft)

3 *Non-direct* type : Involved a degree when is achieved via the interaction with its environment and cannot be commanded. (Pitching motion of a vessel at sea)

Transitional type also exists in some vehicles. For example, when the Space Shuttle operates in space, the craft is described as fully-direct-six because its six degrees can be commanded. However, when the Space Shuttle is in the earth's atmosphere for its return, the fully-direct-six degrees are not longer applicable for many technical reasons.

Game Controllers

First-person shooter (FPS) games generally provide five degrees of freedom: forwards/backwards, slide left/right, up/down (jump/crouch/lie), yaw (turn left/right), and pitch (look up/down). If the game allows leaning control, then some consider it a sixth DoF; however, this may not be completely accurate, as a lean is a limited partial rotation.

The term *6DoF* has sometimes been used to describe games which allow freedom of movement, but do not necessarily meet the full 6DoF criteria. For example, *Dead Space 2*, and to a lesser extent, *Homeworld* and *Zone Of The Enders* allow freedom of movement.

Some examples of true 6DoF games, which allow independent control of all three movement axes and all three rotational axes, include *Shattered Horizon*, the *Descent* franchise, *Retrovirus (PC game)*, *Miner Wars*, *Space Engineers*, *Forsaken and now Overload, from the creators of Descent*. The space MMO *Vendetta Online* also features 6 degrees of freedom.

Motion tracking devices such as TrackIR are used for 6DoF head tracking. This device often finds its places in flight simulators and other vehicle simulators that require looking around the cockpit to locate enemies or simply avoiding accidents in-game.

The acronym 3DoF, meaning movement in the three dimensions but not rotation, is sometimes encountered.

The Razer Hydra, a motion controller for PC, tracks position and rotation of two wired nunchucks, providing six degrees of freedom on each hand.

The SpaceOrb 360 is a 6DOF computer input device released in 1996 originally manufactured and sold by the SpaceTec IMC company (first bought by Labtec, which itself was later bought by Logitech).

Control Theory

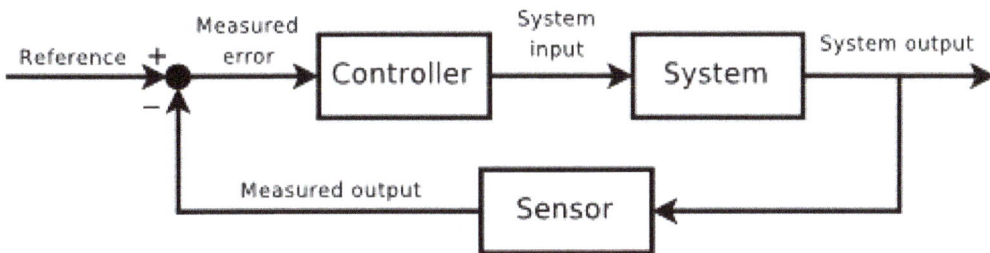

A block diagram of a negative feedback control system. Illustrates the concept of using a feedback loop to control the behavior of a system by comparing its output to a desired value, and applying the difference as an error signal to dynamically change the output so it is closer to the desired output

Control theory is an interdisciplinary branch of engineering and mathematics that deals with the behavior of dynamical systems with inputs, and how their behavior is modified by feedback. The usual objective of control theory is to control a system, often called the *plant*, so its output follows a desired control signal, called the *reference*, which may be a fixed or changing value. To do this a *controller* is designed, which monitors the output and compares it with the reference. The difference between actual and desired output, called the *error* signal, is applied as feedback to the input of the system, to bring the actual output closer to the reference. Some topics studied in control theory are stability (whether the output will converge to the reference value or oscillate about it), controllability and observability.

Extensive use is usually made of a diagrammatic style known as the block diagram. The transfer function, also known as the system function or network function, is a mathematical representation of the relation between the input and output based on the differential equations describing the system.

Although a major application of control theory is in control systems engineering, which deals with the design of process control systems for industry, other applications range far beyond this. As the general theory of feedback systems, control theory is useful wherever feedback occurs. A few examples are in physiology, electronics, climate modeling, machine design, ecosystems, navigation, neural networks, predator–prey interaction, gene expression, and production theory.

Overview

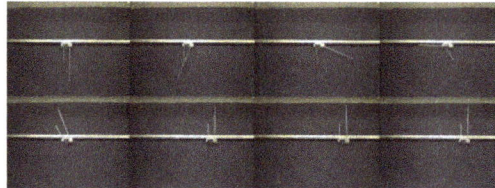

Smooth nonlinear trajectory planning with linear quadratic Gaussian feedback (LQR) control on a dual pendula system.

Control theory is

- a theory that deals with influencing the behavior of dynamical systems

- an interdisciplinary subfield of science, which originated in engineering and mathematics, and evolved into use by the social sciences, such as economics, psychology, sociology, criminology and in the financial system.

Control systems may be thought of as having four functions: measure, compare, compute and correct. These four functions are completed by five elements: detector, transducer, transmitter, controller and final control element. The measuring function is completed by the detector, transducer and transmitter. In practical applications these three elements are typically contained in one unit. A standard example of a measuring unit is a resistance thermometer. The compare and compute functions are completed within the controller, which may be implemented electronically by proportional control, a PI controller, PID controller, bistable, hysteretic control or programmable logic controller. Older controller units have been mechanical, as in a centrifugal governor or a carburetor. The correct function is completed with a final control element. The final control element changes an input or output in the control system that affects the manipulated or controlled variable.

An Example

An example of a control system is a car's cruise control, which is a device designed to maintain vehicle speed at a constant *desired* or *reference* speed provided by the driver. The *controller* is the cruise control, the *plant* is the car, and the *system* is the car and the cruise control. The system output is the car's speed, and the control itself is the engine's throttle position which determines how much power the engine delivers.

A primitive way to implement cruise control is simply to lock the throttle position when the driver engages cruise control. However, if the cruise control is engaged on a stretch of flat road, then the car will travel slower going uphill and faster when going downhill. This type of controller is called an *open-loop controller* because there is no feedback; no measurement of the system output (the car's speed) is used to alter the control (the throttle position.) As a result, the controller cannot compensate for changes acting on the car, like a change in the slope of the road.

In a *closed-loop control system*, data from a sensor monitoring the car's speed (the system output) enters a controller which continuously subtracts the quantity representing the speed from the reference quantity representing the desired speed. The difference, called the error, determines the throttle position (the control). The result is to match the car's speed to the reference speed (maintain the desired system output). Now, when the car goes uphill, the difference between the input (the sensed speed) and the reference continuously determines the throttle position. As the sensed speed drops below the reference, the difference increases, the throttle opens, and engine power increases, speeding up the vehicle. In this way, the controller dynamically counteracts changes to the car's speed. The central idea of these control systems is the *feedback loop*, the controller affects the system output, which in turn is measured and fed back to the controller.

Classification

Linear Versus Nonlinear Control Theory

The field of control theory can be divided into two branches:

- *Linear control theory* – This applies to systems made of devices which obey the superposition principle, which means roughly that the output is proportional to the input. They are governed by linear differential equations. A major subclass is systems which in addition have parameters which do not change with time, called *linear time invariant* (LTI) systems. These systems are amenable to powerful frequency domain mathematical techniques of great generality, such as the Laplace transform, Fourier transform, Z transform, Bode plot, root locus, and Nyquist stability criterion. These lead to a description of the system using terms like bandwidth, frequency response, eigenvalues, gain, resonant frequencies, poles, and zeros, which give solutions for system response and design techniques for most systems of interest.

- *Nonlinear control theory* – This covers a wider class of systems that do not obey the superposition principle, and applies to more real-world systems, because all real control systems are nonlinear. These systems are often governed by nonlinear differential equations. The few mathematical techniques which have been developed to handle them are more difficult and much less general, often applying only to narrow categories of systems. These include limit cycle theory, Poincaré maps, Lyapunov stability theorem, and describing functions. Nonlinear systems are often analyzed using numerical methods on computers, for example by simulating their operation using a simulation language. If only solutions near a stable point are of interest, nonlinear systems can often be linearized by approximating them by a linear system using perturbation theory, and linear techniques can be used.

Frequency Domain Versus Time Domain

Mathematical techniques for analyzing and designing control systems fall into two different categories:

- *Frequency domain* – In this type the values of the state variables, the mathematical variables representing the system's input, output and feedback are represented as functions of frequency. The input signal and the system's transfer function are converted from time functions to functions of frequency by a transform such as the Fourier transform, Laplace transform, or Z transform. The advantage of this technique is that it results in a simplification of the mathematics; the *differential equations* that represent the system are replaced by *algebraic equations* in the frequency domain which are much simpler to solve. However, frequency domain techniques can only be used with linear systems, as mentioned above.

- *Time-domain state space representation* – In this type the values of the state variables are represented as functions of time. With this model the system being analyzed is represented by one or more differential equations. Since frequency domain techniques are limited to linear systems, time domain is widely used to analyze real-world nonlinear systems. Although these are more difficult to solve, modern computer simulation techniques such as simulation languages have made their analysis routine.

SISO vs MIMO

Control systems can be divided into different categories depending on the number of inputs and outputs.

- Single-input single-output (SISO) – This is the simplest and most common type, in which one output is controlled by one control signal. Examples are the cruise control example above, or an audio system, in which the control input is the input audio signal and the output is the sound waves from the speaker.

- Multiple-input multiple-output (MIMO) – These are found in more complicated systems. For example, modern large telescopes such as the Keck and MMT have mirrors composed of many separate segments each controlled by an actuator. The shape of the entire mirror is constantly adjusted by a MIMO active optics control system using input from multiple sensors at the focal plane, to compensate for changes in the mirror shape due to thermal expansion, contraction, stresses as it is rotated and distortion of the wavefront due to turbulence in the atmosphere. Complicated systems such as nuclear reactors and human cells are simulated by computer as large MIMO control systems.

History

Although control systems of various types date back to antiquity, a more formal analysis of the field began with a dynamics analysis of the centrifugal governor, conducted by the physicist James Clerk Maxwell in 1868, entitled *On Governors*. This described and analyzed the phenomenon of self-oscillation, in which lags in the system may lead to overcompensation and unstable behavior. This generated a flurry of interest in the topic, during which Maxwell's classmate, Edward John Routh, abstracted Maxwell's results for the general class of linear systems. Independently, Adolf Hurwitz analyzed system stability using differential equations in 1877, resulting in what is now known as the Routh–Hurwitz theorem.

Centrifugal governor in a Boulton & Watt engine of 1788

A notable application of dynamic control was in the area of manned flight. The Wright brothers made their first successful test flights on December 17, 1903 and were distinguished by their ability to control their flights for substantial periods (more so than the ability to produce lift from an airfoil, which was known). Continuous, reliable control of the airplane was necessary for flights lasting longer than a few seconds.

By World War II, control theory was an important part of fire-control systems, guidance systems and electronics.

Sometimes, mechanical methods are used to improve the stability of systems. For example, ship stabilizers are fins mounted beneath the waterline and emerging laterally. In contemporary vessels, they may be gyroscopically controlled active fins, which have the capacity to change their angle of attack to counteract roll caused by wind or waves acting on the ship.

The Sidewinder missile uses small control surfaces placed at the rear of the missile with spinning disks on their outer surfaces and these are known as rollerons. Airflow over the disks spins them to a high speed. If the missile starts to roll, the gyroscopic force of the disks drives the control surface into the airflow, cancelling the motion. Thus, the Sidewinder team replaced a potentially complex control system with a simple mechanical solution.

The Space Race also depended on accurate spacecraft control, and control theory has also seen an increasing use in fields such as economics.

People in Systems and Control

Many active and historical figures made significant contribution to control theory including

- Pierre-Simon Laplace (1749–1827) invented the Z-transform in his work on probability theory, now used to solve discrete-time control theory problems. The Z-transform is a discrete-time equivalent of the Laplace transform which is named after him.

- Alexander Lyapunov (1857–1918) in the 1890s marks the beginning of stability theory.

- Harold S. Black (1898–1983), invented the concept of negative feedback amplifiers in 1927. He managed to develop stable negative feedback amplifiers in the 1930s.

- Harry Nyquist (1889–1976) developed the Nyquist stability criterion for feedback systems in the 1930s.

- Richard Bellman (1920–1984) developed dynamic programming since the 1940s.

- Andrey Kolmogorov (1903–1987) co-developed the Wiener–Kolmogorov filter in 1941.

- Norbert Wiener (1894–1964) co-developed the Wiener–Kolmogorov filter and coined the term cybernetics in the 1940s.

- John R. Ragazzini (1912–1988) introduced digital control and the use of Z-transform in control theory (invented by Laplace) in the 1950s.

- Lev Pontryagin (1908–1988) introduced the maximum principle and the bang-bang principle.

- Pierre-Louis Lions (1956) developed viscosity solutions into stochastic control and optimal control methods.

Classical Control Theory

To overcome the limitations of the open-loop controller, control theory introduces feedback. A closed-loop controller uses feedback to control states or outputs of a dynamical system. Its name comes from the information path in the system: process inputs (e.g., voltage applied to an electric motor) have an effect on the process outputs (e.g., speed or torque of the motor), which is measured with sensors and processed by the controller; the result (the control signal) is "fed back" as input to the process, closing the loop.

Closed-loop controllers have the following advantages over open-loop controllers:

- disturbance rejection (such as hills in the cruise control example above)

- guaranteed performance even with model uncertainties, when the model structure does not match perfectly the real process and the model parameters are not exact

- unstable processes can be stabilized

- reduced sensitivity to parameter variations

- improved reference tracking performance

In some systems, closed-loop and open-loop control are used simultaneously. In such systems, the open-loop control is termed feedforward and serves to further improve reference tracking performance.

A common closed-loop controller architecture is the PID controller.

Closed-loop Transfer Function

The output of the system $y(t)$ is fed back through a sensor measurement F to a comparison with the reference value $r(t)$. The controller C then takes the error e (difference) between the reference and the output to change the inputs u to the system under control P. This is shown in the figure. This kind of controller is a closed-loop controller or feedback controller.

This is called a single-input-single-output (*SISO*) control system; *MIMO* (i.e., Multi-

Input-Multi-Output) systems, with more than one input/output, are common. In such cases variables are represented through vectors instead of simple scalar values. For some distributed parameter systems the vectors may be infinite-dimensional (typically functions).

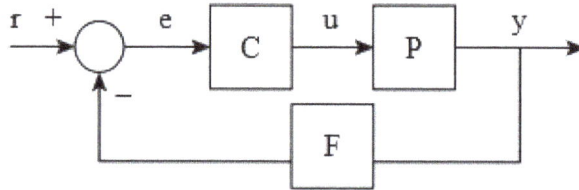

If we assume the controller C, the plant P, and the sensor F are linear and time-invariant (i.e., elements of their transfer function $C(s)$, $P(s)$, and $F(s)$ do not depend on time), the systems above can be analysed using the Laplace transform on the variables. This gives the following relations:

$$Y(s) = P(s)U(s)$$

$$U(s) = C(s)E(s)$$

$$E(s) = R(s) - F(s)Y(s).$$

Solving for $Y(s)$ in terms of $R(s)$ gives

$$Y(s) = \left(\frac{P(s)C(s)}{1 \quad F(s)P(s)C(s)} \right) R(s) = H(s)R(s).$$

The $H(s) = \dfrac{P(s)C(s)}{1 + F(s)P(s)C(s)}$ is referred to as the *closed-loop transfer function* of the system. The numerator is the forward (open-loop) gain from r to y, and the denominator is one plus the gain in going around the feedback loop, the so-called loop gain. If $|P(s)C(s)| \gg 1$, i.e., it has a large norm with each value of s, and if $|F(s)| \approx 1$, then $Y(s)$ is approximately equal to $R(s)$ and the output closely tracks the reference input.

PID Controller

The PID controller is probably the most-used feedback control design. *PID* is an initialism for *Proportional-Integral-Derivative*, referring to the three terms operating on the error signal to produce a control signal. If $u(t)$ is the control signal sent to the system, $y(t)$ is the measured output and $r(t)$ is the desired output, and tracking error $e(t) = r(t) - y(t)$, a PID controller has the general form

$$u(t) = K_p e(t) + K_I \int e(t)\mathrm{d}t + K_D \frac{\mathrm{d}}{\mathrm{d}t} e(t).$$

The desired closed loop dynamics is obtained by adjusting the three parameters K_P

, K_I and K_D, often iteratively by "tuning" and without specific knowledge of a plant model. Stability can often be ensured using only the proportional term. The integral term permits the rejection of a step disturbance (often a striking specification in process control). The derivative term is used to provide damping or shaping of the response. PID controllers are the most well established class of control systems: however, they cannot be used in several more complicated cases, especially if MIMO systems are considered.

Applying Laplace transformation results in the transformed PID controller equation

$$u(s) = K_P e(s) + K_I \frac{1}{s} e(s) + K_D s e(s)$$

$$u(s) = \left(K_P + K_I \frac{1}{s} + K_D s \right) e(s)$$

with the PID controller transfer function

$$C(s) = \left(K_P + K_I \frac{1}{s} + K_D s \right).$$

There exists a nice example of the closed-loop system discussed above. If we take

PID controller transfer function in series form

$$C(s) = K \left(1 + \frac{1}{s T_i} \right)(1 + s T_d)$$

1st order filter in feedback loop

$$F(s) = \frac{1}{1 + s T_f}$$

linear actuator with filtered input

$$P(s) = \frac{A}{1 + s T_p}, \quad A = \text{const}$$

and insert all this into expression for closed-loop transfer function H(s), then tuning is very easy: simply put

$$K = \frac{1}{A}, T_i = T_f, T_d = T_p$$

and get H(s) = 1 identically.

For practical PID controllers, a pure differentiator is neither physically realisable nor desirable due to amplification of noise and resonant modes in the system. Therefore, a phase-lead compensator type approach is used instead, or a differentiator with low-pass roll-off.

Modern Control Theory

In contrast to the frequency domain analysis of the classical control theory, modern control theory utilizes the time-domain state space representation, a mathematical model of a physical system as a set of input, output and state variables related by first-order differential equations. To abstract from the number of inputs, outputs and states, the variables are expressed as vectors and the differential and algebraic equations are written in matrix form (the latter only being possible when the dynamical system is linear). The state space representation (also known as the "time-domain approach") provides a convenient and compact way to model and analyze systems with multiple inputs and outputs. With inputs and outputs, we would otherwise have to write down Laplace transforms to encode all the information about a system. Unlike the frequency domain approach, the use of the state-space representation is not limited to systems with linear components and zero initial conditions. "State space" refers to the space whose axes are the state variables. The state of the system can be represented as a point within that space.

Topics in Control Theory

Stability

The *stability* of a general dynamical system with no input can be described with Lyapunov stability criteria.

- A linear system is called bounded-input bounded-output (BIBO) stable if its output will stay bounded for any bounded input.

- Stability for nonlinear systems that take an input is input-to-state stability (ISS), which combines Lyapunov stability and a notion similar to BIBO stability.

For simplicity, the following descriptions focus on continuous-time and discrete-time linear systems.

Mathematically, this means that for a causal linear system to be stable all of the poles of its transfer function must have negative-real values, i.e. the real part of each pole must be less than zero. Practically speaking, stability requires that the transfer function complex poles reside

- in the open left half of the complex plane for continuous time, when the Laplace transform is used to obtain the transfer function.

- inside the unit circle for discrete time, when the Z-transform is used.

The difference between the two cases is simply due to the traditional method of plotting continuous time versus discrete time transfer functions. The continuous Laplace transform is in Cartesian coordinates where the x axis is the real axis and the discrete Z-transform is in circular coordinates where p axis is the real axis.

When the appropriate conditions above are satisfied a system is said to be asymptotically stable; the variables of an asymptotically stable control system always decrease from their initial value and do not show permanent oscillations. Permanent oscillations occur when a pole has a real part exactly equal to zero (in the continuous time case) or a modulus equal to one (in the discrete time case). If a simply stable system response neither decays nor grows over time, and has no oscillations, it is marginally stable; in this case the system transfer function has non-repeated poles at complex plane origin (i.e. their real and complex component is zero in the continuous time case). Oscillations are present when poles with real part equal to zero have an imaginary part not equal to zero.

If a system in question has an impulse response of

$$x[n] = 0.5^n u[n]$$

then the Z-transform (see this example), is given by

$$X(z) = \frac{1}{1 - 0.5z^{-1}}$$

which has a pole in $z = 0.5$ (zero imaginary part). This system is BIBO (asymptotically) stable since the pole is *inside* the unit circle.

However, if the impulse response was

$$x[n] = 1.5^n u[n]$$

then the Z-transform is

$$X(z) = \frac{1}{1 - 1.5z^{-1}}$$

which has a pole $z = 1.5$ and is not BIBO stable since the pole has a modulus strictly greater than one.

Numerous tools exist for the analysis of the poles of a system. These include graphical systems like the root locus, Bode plots or the Nyquist plots.

Mechanical changes can make equipment (and control systems) more stable. Sailors add ballast to improve the stability of ships. Cruise ships use antiroll fins that extend transversely from the side of the ship for perhaps 30 feet (10 m) and are continuously rotated about their axes to develop forces that oppose the roll.

Controllability and Observability

Controllability and observability are main issues in the analysis of a system before deciding the best control strategy to be applied, or whether it is even possible to control or stabilize the system. Controllability is related to the possibility of forcing the system into a particular state by using an appropriate control signal. If a state is not controllable, then no signal will ever be able to control the state. If a state is not controllable, but its dynamics are stable, then the state is termed *stabilizable*. Observability instead is related to the possibility of *observing*, through output measurements, the state of a system. If a state is not observable, the controller will never be able to determine the behaviour of an unobservable state and hence cannot use it to stabilize the system. However, similar to the stabilizability condition above, if a state cannot be observed it might still be detectable.

From a geometrical point of view, looking at the states of each variable of the system to be controlled, every "bad" state of these variables must be controllable and observable to ensure a good behaviour in the closed-loop system. That is, if one of the eigenvalues of the system is not both controllable and observable, this part of the dynamics will remain untouched in the closed-loop system. If such an eigenvalue is not stable, the dynamics of this eigenvalue will be present in the closed-loop system which therefore will be unstable. Unobservable poles are not present in the transfer function realization of a state-space representation, which is why sometimes the latter is preferred in dynamical systems analysis.

Solutions to problems of uncontrollable or unobservable system include adding actuators and sensors.

Control Specification

Several different control strategies have been devised in the past years. These vary from extremely general ones (PID controller), to others devoted to very particular classes of systems (especially robotics or aircraft cruise control).

A control problem can have several specifications. Stability, of course, is always present. The controller must ensure that the closed-loop system is stable, regardless of the open-loop stability. A poor choice of controller can even worsen the stability of the open-loop system, which must normally be avoided. Sometimes it would be desired to obtain particular dynamics in the closed loop: i.e. that the poles have $Re[\lambda] < -\bar{\lambda}$, where $\bar{\lambda}$ is a fixed value strictly greater than zero, instead of simply asking that $Re[\lambda] < 0$.

Another typical specification is the rejection of a step disturbance; including an integrator in the open-loop chain (i.e. directly before the system under control) easily achieves this. Other classes of disturbances need different types of sub-systems to be included.

Other "classical" control theory specifications regard the time-response of the closed-loop system. These include the rise time (the time needed by the control system to reach the desired value after a perturbation), peak overshoot (the highest value reached by the response before reaching the desired value) and others (settling time, quarter-decay). Frequency domain specifications are usually related to robustness.

Modern performance assessments use some variation of integrated tracking error (IAE,ISA,CQI).

Model Identification and Robustness

A control system must always have some robustness property. A robust controller is such that its properties do not change much if applied to a system slightly different from the mathematical one used for its synthesis. This specification is important, as no real physical system truly behaves like the series of differential equations used to represent it mathematically. Typically a simpler mathematical model is chosen in order to simplify calculations, otherwise the true system dynamics can be so complicated that a complete model is impossible.

System Identification

The process of determining the equations that govern the model's dynamics is called system identification. This can be done off-line: for example, executing a series of measures from which to calculate an approximated mathematical model, typically its transfer function or matrix. Such identification from the output, however, cannot take account of unobservable dynamics. Sometimes the model is built directly starting from known physical equations, for example, in the case of a mass-spring-damper system we know that $m\ddot{x}(t) = -Kx(t) - B\dot{x}(t)..$ Even assuming that a "complete" model is used in designing the controller, all the parameters included in these equations (called "nominal parameters") are never known with absolute precision; the control system will have to behave correctly even when connected to physical system with true parameter values away from nominal.

Some advanced control techniques include an "on-line" identification process. The parameters of the model are calculated ("identified") while the controller itself is running. In this way, if a drastic variation of the parameters ensues, for example, if the robot's arm releases a weight, the controller will adjust itself consequently in order to ensure the correct performance.

Analysis

Analysis of the robustness of a SISO (single input single output) control system can be performed in the frequency domain, considering the system's transfer function and using Nyquist and Bode diagrams. Topics include gain and phase margin and amplitude margin. For MIMO (multi input multi output) and, in general, more complicated control systems one must consider the theoretical results devised for each control technique. I.e., if particular robustness qualities are needed, the engineer must shift his attention to a control technique by including them in its properties.

Constraints

A particular robustness issue is the requirement for a control system to perform properly in the presence of input and state constraints. In the physical world every signal is limited. It could happen that a controller will send control signals that cannot be followed by the physical system, for example, trying to rotate a valve at excessive speed. This can produce undesired behavior of the closed-loop system, or even damage or break actuators or other subsystems. Specific control techniques are available to solve the problem: model predictive control, and anti-wind up systems. The latter consists of an additional control block that ensures that the control signal never exceeds a given threshold.

System Classifications

Linear Systems Control

For MIMO systems, pole placement can be performed mathematically using a state space representation of the open-loop system and calculating a feedback matrix assigning poles in the desired positions. In complicated systems this can require computer-assisted calculation capabilities, and cannot always ensure robustness. Furthermore, all system states are not in general measured and so observers must be included and incorporated in pole placement design.

Nonlinear Systems Control

Processes in industries like robotics and the aerospace industry typically have strong nonlinear dynamics. In control theory it is sometimes possible to linearize such classes of systems and apply linear techniques, but in many cases it can be necessary to devise from scratch theories permitting control of nonlinear systems. These, e.g., feedback linearization, backstepping, sliding mode control, trajectory linearization control normally take advantage of results based on Lyapunov's theory. Differential geometry has been widely used as a tool for generalizing well-known linear control concepts to the non-linear case, as well as showing the subtleties that make it a more challenging problem. Control theory has also been used to decipher the neural mechanism that directs cognitive states.

Decentralized Systems Control

When the system is controlled by multiple controllers, the problem is one of decentralized control. Decentralization is helpful in many ways, for instance, it helps control systems to operate over a larger geographical area. The agents in decentralized control systems can interact using communication channels and coordinate their actions.

Deterministic and Stochastic Systems Control

A stochastic control problem is one in which the evolution of the state variables is subjected to random shocks from outside the system. A deterministic control problem is not subject to external random shocks.

Main Control Strategies

Every control system must guarantee first the stability of the closed-loop behavior. For linear systems, this can be obtained by directly placing the poles. Non-linear control systems use specific theories (normally based on Aleksandr Lyapunov's Theory) to ensure stability without regard to the inner dynamics of the system. The possibility to fulfill different specifications varies from the model considered and the control strategy chosen.

List of the Main Control Techniques

- Adaptive control uses on-line identification of the process parameters, or modification of controller gains, thereby obtaining strong robustness properties. Adaptive controls were applied for the first time in the aerospace industry in the 1950s, and have found particular success in that field.

- A hierarchical control system is a type of control system in which a set of devices and governing software is arranged in a hierarchical tree. When the links in the tree are implemented by a computer network, then that hierarchical control system is also a form of networked control system.

- Intelligent control uses various AI computing approaches like neural networks, Bayesian probability, fuzzy logic, machine learning, evolutionary computation and genetic algorithms to control a dynamic system.

- Optimal control is a particular control technique in which the control signal optimizes a certain "cost index": for example, in the case of a satellite, the jet thrusts needed to bring it to desired trajectory that consume the least amount of fuel. Two optimal control design methods have been widely used in industrial applications, as it has been shown they can guarantee closed-loop stability. These are Model Predictive Control (MPC) and linear-quadratic-Gaussian control (LQG). The first can more explicitly take into account constraints on the signals in the system, which is an important feature in many industrial processes. However, the "optimal control" structure in MPC is only a means to achieve such a result, as it does not optimize a true performance index of the closed-loop control system. Together with PID controllers, MPC systems are the most widely used control technique in process control.

- Robust control deals explicitly with uncertainty in its approach to controller design. Controllers designed using *robust control* methods tend to be able to cope with small differences between the true system and the nominal model used for design. The early methods of Bode and others were fairly robust; the state-space methods invented in the 1960s and 1970s were sometimes found to lack robustness. Examples of modern robust control techniques include H-infinity loop-shaping developed by Duncan McFarlane and Keith Glover of Cambridge University, United Kingdom and Sliding mode control (SMC) developed by Vadim Utkin. Robust methods aim to achieve robust performance and/or stability in the presence of small modeling errors.

- Stochastic control deals with control design with uncertainty in the model. In typical stochastic control problems, it is assumed that there exist random noise and disturbances in the model and the controller, and the control design must take into account these random deviations.

- Energy-shaping control view the plant and the controller as energy-transformation devices. The control strategy is formulated in terms of interconnection (in a power-preserving manner) in order to achieve a desired behavior.

- Self-organized criticality control may be defined as attempts to interfere in the processes by which the self-organized system dissipates energy.

Gimbal Lock

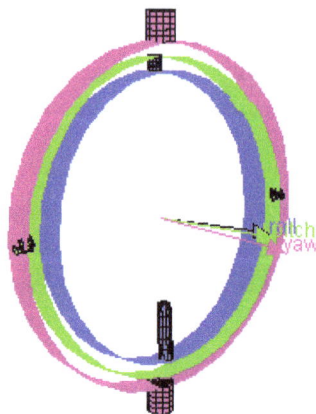

Gimbal with 3 axes of rotation. A set of three gimbals mounted together to allow three degrees of freedom: roll, pitch and yaw. When two gimbals rotate around the same axis, the system loses one degree of freedom.

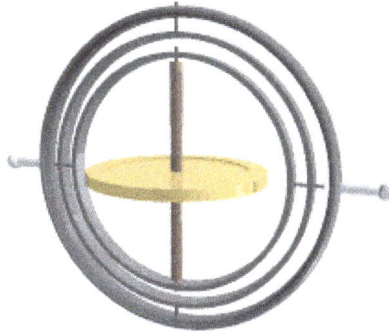

Adding a fourth rotational axis can solve the problem of gimbal lock, but it requires the outermost ring to be actively driven so that it stays 90 degrees out of alignment with the innermost axis (the flywheel shaft). Without active driving of the outermost ring, all four axes can become aligned in a plane as shown above, again leading to gimbal lock and inability to roll.

Gimbal lock is the loss of one degree of freedom in a three-dimensional, three-gimbal mechanism that occurs when the axes of two of the three gimbals are driven into a parallel configuration, "locking" the system into rotation in a degenerate two-dimensional space.

The word *lock* is misleading: no gimbal is restrained. All three gimbals can still rotate freely about their respective axes of suspension. Nevertheless, because of the parallel orientation of two of the gimbals' axes there is no gimbal available to accommodate rotation along one axis.

Gimbals

A gimbal is a ring that is suspended so it can rotate about an axis. Gimbals are typically nested one within another to accommodate rotation about multiple axes.

They appear in gyroscopes and in inertial measurement units to allow the inner gimbal's orientation to remain fixed while the outer gimbal suspension assumes any orientation. In compasses and flywheel energy storage mechanisms they allow objects to remain upright. They are used to orient thrusters on rockets.

Some coordinate systems in mathematics behave as if there were real gimbals used to measure the angles, notably Euler angles.

For cases of three or fewer nested gimbals, gimbal lock inevitably occurs at some point in the system due to properties of covering spaces (described below).

Gimbal Lock in Mechanical Engineering

While only two specific orientations produce exact gimbal lock, practical mechanical gimbals encounter difficulties near those orientations. When a set of gimbals are close to the locked configuration, small rotations of the gimbal platform require large motions of the surrounding gimbals. Although the ratio is infinite only at the point of gimbal lock, the practical speed and acceleration limits of the gimbals limit the motion of the platform close to that point.

Gimbal Lock in Two Dimensions

Gimbal lock can occur in gimbal systems with two degrees of freedom such as a theodolite with rotations about an azimuth and elevation in two dimensions. These systems can gimbal lock at zenith and nadir, because at those points azimuth is not well-defined, and rotation in the azimuth direction does not change the direction the theodolite is pointing.

Consider tracking a helicopter flying towards the theodolite from the horizon. The theodolite is a telescope mounted on a tripod so that it can move in azimuth and elevation to track the helicopter. The helicopter flies towards the theodolite and is tracked by the telescope in elevation and azimuth. The helicopter flies immediately above the tripod (i.e. it is at zenith) when it changes direction and flies at 90 degrees to its previous course. The telescope cannot track this maneuver without a discontinuous jump in one or both of the gimbal orientations. There is no continuous motion that allows it to follow the target. It is in gimbal lock. So there is an infinity of directions around zenith for which the telescope cannot continuously track all movements of a target. Note that even if the helicopter does not pass through zenith, but only *near* zenith, so that gimbal lock does not occur, the system must still move exceptionally rapidly to track it, as it rapidly passes from one bearing to the other. The closer to zenith the nearest point is, the faster this must be done, and if it actually goes through zenith, the limit of these "increasingly rapid" movements becomes *infinitely* fast, namely discontinuous.

To recover from gimbal lock the user has to go around the zenith – explicitly: reduce the elevation, change the azimuth to match the azimuth of the target, then change the elevation to match the target.

Mathematically, this corresponds to the fact that spherical coordinates do not define a coordinate chart on the sphere at zenith and nadir. Alternatively, the corresponding map $T^2 \to S^2$ from the torus T^2 to the sphere S^2 (given by the point with given azimuth and elevation) is not a covering map at these points.

Gimbal Lock in Three Dimensions

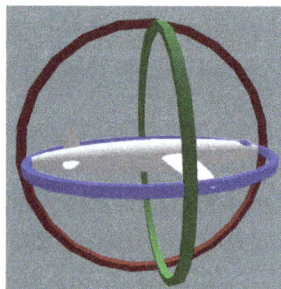

Normal situation: the three gimbals are independent

Gimbal lock: two out of the three gimbals are in the same plane, one degree of freedom is lost

Consider a case of a level sensing platform on an aircraft flying due north with its three gimbal axes mutually perpendicular (i.e., roll, pitch and yaw angles each zero). If the aircraft pitches up 90 degrees, the aircraft and platform's yaw axis gimbal becomes parallel to the roll axis gimbal, and changes about yaw can no longer be compensated for.

Solutions

This problem may be overcome by use of a fourth gimbal, intelligently driven by a motor so as to maintain a large angle between roll and yaw gimbal axes. Another solution is to rotate one or more of the gimbals to an arbitrary position when gimbal lock is detected and thus reset the device.

Modern practice is to avoid the use of gimbals entirely. In the context of inertial navigation systems, that can be done by mounting the inertial sensors directly to the body of the vehicle (this is called a strapdown system) and integrating sensed rotation and acceleration digitally using quaternion methods to derive vehicle orientation and velocity. Another way to replace gimbals is to use fluid bearings or a flotation chamber.

Gimbal Lock on Apollo 11

A well-known gimbal lock incident happened in the Apollo 11 Moon mission. On this spacecraft, a set of gimbals was used on an inertial measurement unit (IMU). The engineers were aware of the gimbal lock problem but had declined to use a fourth gimbal. Some of the reasoning behind this decision is apparent from the following quote:

"The advantages of the redundant gimbal seem to be outweighed by the equipment simplicity, size advantages, and corresponding implied reliability of the direct three degree of freedom unit."

— *David Hoag, Apollo Lunar Surface Journal*

They preferred an alternate solution using an indicator that would be triggered when near to 85 degrees pitch.

"Near that point, in a closed stabilization loop, the torque motors could theoretically be commanded to flip the gimbal 180 degrees instantaneously. Instead, in the LM, the computer flashed a 'gimbal lock' warning at 70 degrees and froze the IMU at 85 degrees"

— Paul Fjeld, Apollo Lunar Surface Journal

Rather than try to drive the gimbals faster than they could go, the system simply gave up and froze the platform. From this point, the spacecraft would have to be manually moved away from the gimbal lock position, and the platform would have to be manually realigned using the stars as a reference.

After the Lunar Module had landed, Mike Collins aboard the Command Module joked "How about sending me a fourth gimbal for Christmas?"

Robotics

Industrial robot operating in a foundry.

In robotics, gimbal lock is commonly referred to as "wrist flip", due to the use of a "triple-roll wrist" in robotic arms, where three axes of the wrist, controlling yaw, pitch, and roll, all pass through a common point.

An example of a wrist flip, also called a wrist singularity, is when the path through which the robot is traveling causes the first and third axes of the robot's wrist to line up. The second wrist axis then attempts to spin 180° in zero time to maintain the orientation of the end effector. The result of a singularity can be quite dramatic and can have adverse effects on the robot arm, the end effector, and the process.

The importance of avoiding singularities in robotics has led the American National Standard for Industrial Robots and Robot Systems — Safety Requirements to define it as "a condition caused by the collinear alignment of two or more robot axes resulting in unpredictable robot motion and velocities".

Gimbal Lock in Applied Mathematics

The problem of gimbal lock appears when one uses Euler angles in applied mathematics; developers of 3D computer programs, such as 3D modeling, embedded navigation systems, and video games must take care to avoid it.

In formal language, gimbal lock occurs because the map from Euler angles to rotations (topologically, from the 3-torus T^3 to the real projective space RP³) is not a covering map – it is not a local homeomorphism at every point, and thus at some points the rank (degrees of freedom) must drop below 3, at which point gimbal lock occurs. Euler angles provide a means for giving a numerical description of any rotation in three-dimensional space using three numbers, but not only is this description not unique, but there are some points where not every change in the target space (rotations) can be realized by a change in the source space (Euler angles). This is a topological constraint – there is no covering map from the 3-torus to the 3-dimensional real projective space; the only (non-trivial) covering map is from the 3-sphere, as in the use of quaternions.

To make a comparison, all the translations can be described using three numbers x, y, and z, as the succession of three consecutive linear movements along three perpendicular axes X, Y and Z axes. The same holds true for rotations: all the rotations can be described using three numbers α, β, and γ, as the succession of three rotational movements around three axes that are perpendicular one to the next. This similarity between linear coordinates and angular coordinates makes Euler angles very intuitive, but unfortunately they suffer from the gimbal lock problem.

Loss of a Degree of Freedom with Euler Angles

A rotation in 3D space can be represented numerically with matrices in several ways. One of these representations is:

$$R = \begin{bmatrix} 1 & 0 & 0 \\ 0 & \cos\alpha & -\sin\alpha \\ 0 & \sin\alpha & \cos\alpha \end{bmatrix} \begin{bmatrix} \cos\beta & 0 & \sin\beta \\ 0 & 1 & 0 \\ -\sin\beta & 0 & \cos\beta \end{bmatrix} \begin{bmatrix} \cos\gamma & -\sin\gamma & 0 \\ \sin\gamma & \cos\gamma & 0 \\ 0 & 0 & 1 \end{bmatrix}$$

An example worth examining happens when $\beta = \frac{\pi}{2}$. Knowing that $\cos\frac{\pi}{2} = 0$ and $\sin\frac{\pi}{2} = 1$, the above expression becomes equal to:

$$R = \begin{bmatrix} 1 & 0 & 0 \\ 0 & \cos\alpha & -\sin\alpha \\ 0 & \sin\alpha & \cos\alpha \end{bmatrix} \begin{bmatrix} 0 & 0 & 1 \\ 0 & 1 & 0 \\ -1 & 0 & 0 \end{bmatrix} \begin{bmatrix} \cos\gamma & -\sin\gamma & 0 \\ \sin\gamma & \cos\gamma & 0 \\ 0 & 0 & 1 \end{bmatrix}$$

Carrying out matrix multiplication:

$$R = \begin{bmatrix} 0 & 0 & 1 \\ \sin\alpha & \cos\alpha & 0 \\ -\cos\alpha & \sin\alpha & 0 \end{bmatrix} \begin{bmatrix} \cos\gamma & -\sin\gamma & 0 \\ \sin\gamma & \cos\gamma & 0 \\ 0 & 0 & 1 \end{bmatrix} = \begin{bmatrix} 0 & 0 & 1 \\ \sin\alpha\cos\gamma + \cos\alpha\sin\gamma & -\sin\alpha\sin\gamma + \cos\alpha\cos\gamma & 0 \\ -\cos\alpha\cos\gamma + \sin\alpha\sin\gamma & \cos\alpha\sin\gamma + \sin\alpha\cos\gamma & 0 \end{bmatrix}$$

And finally using the trigonometry formulas:

$$R = \begin{bmatrix} 0 & 0 & 1 \\ \sin(\alpha+\gamma) & \cos(\alpha+\gamma) & 0 \\ -\cos(\alpha+\gamma) & \sin(\alpha+\gamma) & 0 \end{bmatrix}$$

Changing the values of α and γ in the above matrix has the same effects: the rotation angle $\alpha + \gamma$ changes, but the rotation axis remains in the Z direction: the last column and the first row in the matrix won't change. The only solution for α and γ to recover different roles is to change β.

It is possible to imagine an airplane rotated by the above-mentioned Euler angles using the X-Y-Z convention. In this case, the first angle - α is the pitch. Yaw is then set to $\dfrac{\pi}{2}$ and the final rotation - by γ - is again the airplane's pitch. Because of gimbal lock, it has lost one of the degrees of freedom - in this case the ability to roll.

It is also possible to choose another convention for representing a rotation with a matrix using Euler angles than the X-Y-Z convention above, and also choose other variation intervals for the angles, but in the end there is always at least one value for which a degree of freedom is lost.

The gimbal lock problem does not make Euler angles "invalid" (they always serve as a well-defined coordinate system), but it makes them unsuited for some practical applications.

Alternate Orientation Representation

The cause of gimbal lock is representing an orientation as 3 axial rotations with Euler angles. A potential solution therefore is to represent the orientation in some other way. This could be as a rotation matrix, a quaternion, or a similar orientation representation that treats the orientation as a value rather than 3 separate and related values. Given such a representation, the user stores the orientation as a value. To apply angular changes, the orientation is modified by a delta angle/axis rotation. The resulting orientation must be renormalized to prevent floating-point error from successive transformations from accumulating. For matrices, re-normalizing the result requires converting the matrix into its nearest orthonormal representation. For quaternions, renormalization requires performing quaternion normalization.

Motion Planning

Motion planning (also known as the navigation problem or the piano mover's problem) is a term used in robotics for the process of breaking down a desired movement task into discrete motions that satisfy movement constraints and possibly optimize some aspect of the movement.

For example, consider navigating a mobile robot inside a building to a distant waypoint. It should execute this task while avoiding walls and not falling down stairs. A motion planning algorithm would take a description of these tasks as input, and produce the speed and turning commands sent to the robot's wheels. Motion planning algorithms might address robots with a larger number of joints (e.g., industrial manipulators), more complex tasks (e.g. manipulation of objects), different constraints (e.g., a car that can only drive forward), and uncertainty (e.g. imperfect models of the environment or robot).

Motion planning has several robotics applications, such as autonomy, automation, and robot design in CAD software, as well as applications in other fields, such as animating digital characters, video game artificial intelligence, architectural design, robotic surgery, and the study of biological molecules.

Concepts

Example of a workspace.

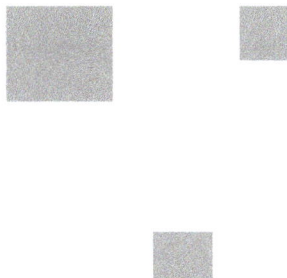

Configuration space of a point-sized robot. White = C_{free}, gray = C_{obs}.

Configuration space for a rectangular translating robot (pictured red). White = C_{free}, gray = C_{obs}, where dark gray = the objects, light gray = configurations where the robot would touch an object or leave the workspace.

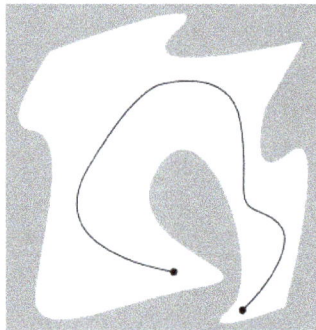

Example of a valid path.

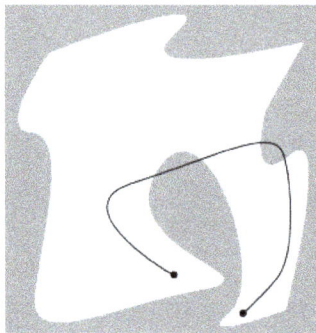

Example of an invalid path.

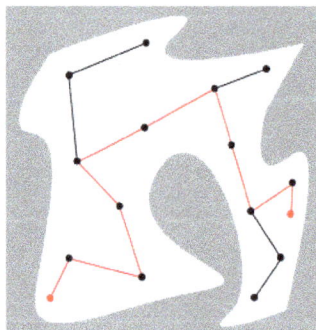

Example of a road map.

A basic motion planning problem is to produce a continuous motion that connects a start configuration S and a goal configuration G, while avoiding collision with known obstacles. The robot and obstacle geometry is described in a 2D or 3D *workspace*, while the motion is represented as a path in (possibly higher-dimensional) configuration space.

Configuration Space

A configuration describes the pose of the robot, and the configuration space C is the set of all possible configurations. For example:

- If the robot is a single point (zero-sized) translating in a 2-dimensional plane (the workspace), C is a plane, and a configuration can be represented using two parameters (x, y).

- If the robot is a 2D shape that can translate and rotate, the workspace is still 2-dimensional. However, C is the special Euclidean group $SE(2) = R^2 SO(2)$ (where SO(2) is the special orthogonal group of 2D rotations), and a configuration can be represented using 3 parameters (x, y, θ).

- If the robot is a solid 3D shape that can translate and rotate, the workspace is 3-dimensional, but C is the special Euclidean group $SE(3) = R^3 SO(3)$, and a configuration requires 6 parameters: (x, y, z) for translation, and Euler angles (α, β, γ).

- If the robot is a fixed-base manipulator with N revolute joints (and no closed-loops), C is N-dimensional.

Free Space

The set of configurations that avoids collision with obstacles is called the free space C_{free}. The complement of C_{free} in C is called the obstacle or forbidden region.

Often, it is prohibitively difficult to explicitly compute the shape of C_{free}. However, testing whether a given configuration is in C_{free} is efficient. First, forward kinematics determine the position of the robot's geometry, and collision detection tests if the robot's geometry collides with the environment's geometry.

Target Space

Target space is a linear subspace of free space which we want robot go there. In global motion planning, target space is observable by robot's sensors. However, in local motion planning, robot cannot observe the target space in some states. To solve problem, robot assume several virtual target space which is located in observable area (around robot). The virtual target space is called sub-goal.

Algorithms

Low-dimensional problems can be solved with grid-based algorithms that overlay a grid on top of configuration space, or geometric algorithms that compute the shape and connectivity of C_{free}.

Exact motion planning for high-dimensional systems under complex constraints is computationally intractable. Potential-field algorithms are efficient, but fall prey to local minima (an exception is the harmonic potential fields). Sampling-based algorithms avoid the problem of local minima, and solve many problems quite quickly. They are unable to determine that no path exists, but they have a probability of failure that decreases to zero as more time is spent.

Sampling-based algorithms are currently considered state-of-the-art for motion planning in high-dimensional spaces, and have been applied to problems which have dozens or even hundreds of dimensions (robotic manipulators, biological molecules, animated digital characters, and legged robots).

Grid-based Search

Grid-based approaches overlay a grid on configuration space, and assume each configuration is identified with a grid point. At each grid point, the robot is allowed to move to adjacent grid points as long as the line between them is completely contained within C_{free} (this is tested with collision detection). This discretizes the set of actions, and search algorithms (like A*) are used to find a path from the start to the goal.

These approaches require setting a grid resolution. Search is faster with coarser grids, but the algorithm will fail to find paths through narrow portions of C_{free}. Furthermore, the number of points on the grid grows exponentially in the configuration space dimension, which make them inappropriate for high-dimensional problems.

Traditional grid-based approaches produce paths whose heading changes are constrained to multiples of a given base angle, often resulting in suboptimal paths. Any-angle path planning approaches find shorter paths by propagating information along grid edges (to search fast) without constraining their paths to grid edges (to find short paths).

Grid-based approaches often need to search repeatedly, for example, when the knowledge of the robot about the configuration space changes or the configuration space itself changes during path following. Incremental heuristic search algorithms replan fast by using experience with the previous similar path-planning problems to speed up their search for the current one.

Interval-based Search

These approaches are similar to grid-based search approaches except that they generate a paving covering entirely the configuration space instead of a grid . The paving is decomposed into two subpavings X^-, X^+ made with boxes such that $X^- \subset C_{\text{free}} \subset X^+$. Characterizing C_{free} amounts to solve a set inversion problem. Interval analysis could thus be used when C_{free} cannot be described by linear inequalities in order to have a guaranteed enclosure.

The robot is thus allowed to move freely in X^-, and cannot go outside X^+. To both subpavings, a neighbor graph is built and paths can be found using algorithms such as Dijkstra or A*. When a path is feasible in X^-, it is also feasible in C_{free}. When no path exists in X^+ from one initial configuration to the goal, we have the guarantee that no feasible path exists in C_{free}. As for the grid-based approach, the interval approach is inappropriate for high-dimensional problems, due to the fact that the number of boxes to be generated grows exponentially with respect to the dimension of configuration space.

An illustration is provided by the three figures on the right where a hook with two degrees of freedom has to move from the left to the right, avoiding two horizontal small segments.

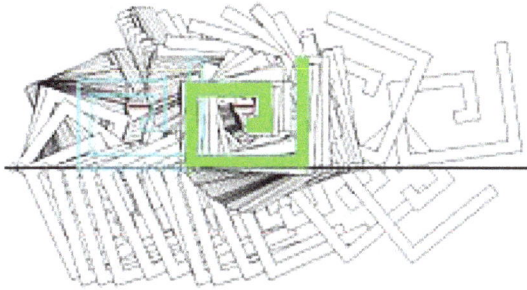

Motion from the initial configuration (blue) to the final configuration of the hook, avoiding the two obstacles (red segments). The left-bottom corner of the hook has to stay on the horizontal line, which makes the hook two degrees of freedom.

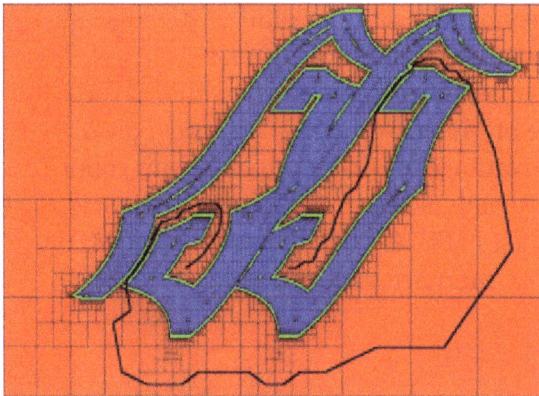

Decomposition with boxes covering the configuration space: The subpaving X^- is the union all red boxes and the subpaving X^+ is the union of red and green boxes. The path corresponds to the motion represented above.

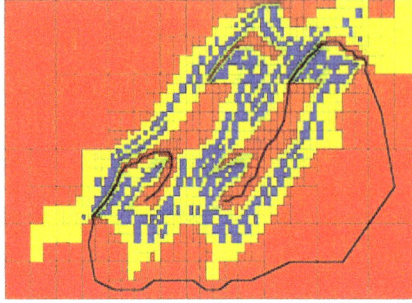

This figure corresponds to the same path as above but obtained with many fewer boxes. The algorithm avoids bisecting boxes in parts of the configuration space that do not influence the final result.

The decomposition with subpavings using interval analysis also makes it possible to characterize the topology of C_{free} such as counting its number of connected components .

Geometric Algorithms

Point robots among polygonal obstacles

- Visibility graph
- Cell decomposition

Translating objects among obstacles

- Minkowski sum

Reward-based Algorithms

Reward-Based Algorithms assume that robot in each state (position and internal state include direction) can choose between different action (motion). However, the result of each action is not definite. In the other word, outcomes (displacement) are partly random and partly under the control of the robot. Robot gets positive reward when it reach to the target and get negative reward if collide with obstacle. These Algorithms try to find a path which maximized cumulative future rewards. Markov decision processes (MDPs) is a popular mathematical framework which is used in many of Reward-Based Algorithms. Advantage of MDPs over other Reward-Based Algorithms is that it generate optimal path. Disadvantage of MDPs is that it limit robot to choose from a finite set of action; Therefore, the path is not smooth (similar to Grid-based approaches). Fuzzy Markov decision processes (FDMPs)is an extension of MDPs which generate smooth path with using an fuzzy inference system .

Artificial Potential Fields

One approach is to treat the robot's configuration as a point (usually electron) in a potential field that combines attraction to the goal, and repulsion from obstacles. The resulting trajectory is output as the path. This approach has advantages in that the trajectory is produced with little computation. However, they can become trapped in local minima of the potential field, and fail to find a path. The Artificial potential fields can be achieved by direct equation similar to electrostatic potential fields or can be drive by set of linguistic rules .

Sampling-based Algorithms

Sampling-based algorithms represent the configuration space with a roadmap of sampled configurations. A basic algorithm samples N configurations in C, and retains those in C_{free} to use as *milestones*. A roadmap is then constructed that connects two milestones P and Q if the line segment PQ is completely in C_{free}. Again, collision detection is used to test inclusion in C_{free}. To find a path that connects S and G, they are added to the roadmap. If a path in the roadmap links S and G, the planner succeeds, and returns that path. If not, the reason is not definitive: either there is no path in C_{free}, or the planner did not sample enough milestones.

These algorithms work well for high-dimensional configuration spaces, because unlike combinatorial algorithms, their running time is not (explicitly) exponentially dependent on the dimension of C. They are also (generally) substantially easier to implement. They are probabilistically complete, meaning the probability that they will produce a solution approaches 1 as more time is spent. However, they cannot determine if no solution exists.

Given basic *visibility* conditions on C_{free}, it has been proven that as the number of configurations N grows higher, the probability that the above algorithm finds a solution approaches 1 exponentially. Visibility is not explicitly dependent on the dimension of C; it is possible to have a high-dimensional space with "good" visibility or a low-dimensional space with "poor" visibility. The experimental success of sample-based methods suggests that most commonly seen spaces have good visibility.

There are many variants of this basic scheme:

- It is typically much faster to only test segments between nearby pairs of milestones, rather than all pairs.

- Nonuniform sampling distributions attempt to place more milestones in areas that improve the connectivity of the roadmap.

- Quasirandom samples typically produce a better covering of configuration space than pseudorandom ones, though some recent work argues that the effect of the source of randomness is minimal compared to the effect of the sampling distribution.

- It is possible to substantially reduce the number of milestones needed to solve a given problem by allowing curved eye sights (for example by crawling on the obstacles that block the way between two milestones).

- If only one or a few planning queries are needed, it is not always necessary to construct a roadmap of the entire space. Tree-growing variants are typically faster for this case (single-query planning). Roadmaps are still useful if many queries are to be made on the same space (multi-query planning)

List of Notable Algorithms

- A*

- D*

- Rapidly-exploring random tree

- Probabilistic roadmap

Completeness and Performance

A motion planner is said to be complete if the planner in finite time either produces a solution or correctly reports that there is none. Most complete algorithms are geometry-based. The performance of a complete planner is assessed by its computational complexity.

Resolution completeness is the property that the planner is guaranteed to find a path if the resolution of an underlying grid is fine enough. Most resolution complete planners are grid-based or interval-based. The computational complexity of resolution complete planners is dependent on the number of points in the underlying grid, which is $O(1/h^d)$, where h is the resolution (the length of one side of a grid cell) and d is the configuration space dimension.

Probabilistic completeness is the property that as more "work" is performed, the probability that the planner fails to find a path, if one exists, asymptotically approaches zero. Several sample-based methods are probabilistically complete. The performance of a probabilistically complete planner is measured by the rate of convergence.

Incomplete planners do not always produce a feasible path when one exists. Sometimes incomplete planners do work well in practice.

Problem Variants

Many algorithms have been developed to handle variants of this basic problem.

Differential Constraints

Holonomic

- Manipulator arms (with dynamics)

Nonholonomic

- Cars

- Unicycles

- Planes

- Acceleration bounded systems

- Moving obstacles (time cannot go backward)

- Bevel-tip steerable needle

- Differential Drive Robots

Optimality Constraints

Hybrid Systems

Hybrid systems are those that mix discrete and continuous behavior. Examples of such systems are:

- Robotic manipulation

- Mechanical assembly

- Legged robot locomotion

- Reconfigurable robots

Uncertainty

- Motion uncertainty

- Missing information

- Active sensing

- Sensorless planning

Shortest Path Problem

In graph theory, the shortest path problem is the problem of finding a path between two vertices (or nodes) in a graph such that the sum of the weights of its constituent edges is minimized.

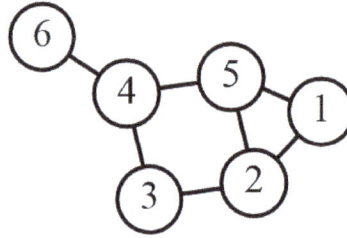

(6, 4, 5, 1) and (6, 4, 3, 2, 1) are both paths between vertices 6 and 1

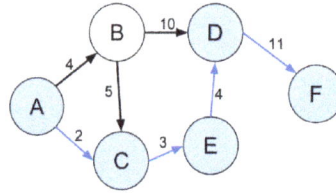

Shortest path (A, C, E, D, F) between vertices A and F in the weighted directed graph

The problem of finding the shortest path between two intersections on a road map (the graph's vertices correspond to intersections and the edges correspond to road segments, each weighted by the length of its road segment) may be modeled by a special case of the shortest path problem in graphs.

Definition

The shortest path problem can be defined for graphs whether undirected, directed, or mixed. It is defined here for undirected graphs; for directed graphs the definition of path requires that consecutive vertices be connected by an appropriate directed edge.

Two vertices are adjacent when they are both incident to a common edge. A path in an undirected graph is a sequence of vertices $P = (v_1, v_2, \ldots, v_n) \in V \times V \times \ldots \times V$ such that v_i is adjacent to v_{i+1} for $1 \leq i < n$. Such a path P is called a path of length $n-1$ from v to v_n. (The v_i are variables; their numbering here relates to their position in the sequence and needs not to relate to any canonical labeling of the vertices.)

Let $e_{i,j}$ be the edge incident to both v_i and v_j. Given a real-valued weight function $f : E \rightarrow \mathbb{R}$, and an undirected (simple) graph G, the shortest path from v to v' is the path $P = (v_1, v_2, \ldots, v_n)$ (where $v_1 = v$ and $v_n = v'$) that over all possible n minimizes the sum $\sum_{i=1}^{n-1} f(e_{i,i+1})$. When each edge in the graph has unit weight or $f : E \rightarrow \{1\}$, this is equivalent to finding the path with fewest edges.

The problem is also sometimes called the single-pair shortest path problem, to distinguish it from the following variations:

- The single-source shortest path problem, in which we have to find shortest paths from a source vertex v to all other vertices in the graph.

- The single-destination shortest path problem, in which we have to find shortest paths from all vertices in the directed graph to a single destination vertex v. This can be reduced to the single-source shortest path problem by reversing the arcs in the directed graph.

- The all-pairs shortest path problem, in which we have to find shortest paths between every pair of vertices v, v' in the graph.

These generalizations have significantly more efficient algorithms than the simplistic approach of running a single-pair shortest path algorithm on all relevant pairs of vertices.

Algorithms

The most important algorithms for solving this problem are:

- Dijkstra's algorithm solves the single-source shortest path problem.

- Bellman–Ford algorithm solves the single-source problem if edge weights may be negative.

- A* search algorithm solves for single pair shortest path using heuristics to try to speed up the search.

- Floyd–Warshall algorithm solves all pairs shortest paths.

- Johnson's algorithm solves all pairs shortest paths, and may be faster than Floyd–Warshall on sparse graphs.

- Viterbi algorithm solves the shortest stochastic path problem with an additional probabilistic weight on each node.

Additional algorithms and associated evaluations may be found in Cherkassky, Goldberg & Radzik (1996).

Single-source Shortest Paths

Undirected Graphs

Weights	Time complexity	Author
R_+	$O(V^2)$	Dijkstra 1959
R_+	$O(E + V \log V)$	Fredman & Tarjan 1984 (Fibonacci heap)
N	$O(E)$	Thorup 1999 (requires constant-time multiplication).

Unweighted Graphs

Algorithm	Time complexity	Author
Breadth-first search	$O(E + V)$	

Directed Acyclic Graphs

An algorithm using topological sorting can solve the single-source shortest path problem in linear time, $\Theta(E + V)$, in weighted DAGs.

Directed Graphs with Nonnegative Weights

The following table is taken from Schrijver (2004). A green background indicates an asymptotically best bound in the table.

Algorithm	Time complexity	Author
	$O(V^2EL)$	Ford 1956
Bellman–Ford algorithm	$O(VE)$	Bellman 1958, Moore 1959
	$O(V^2 \log V)$	Dantzig 1958, Dantzig 1960, Minty (cf. Pollack & Wiebenson 1960), Whiting & Hillier 1960
Dijkstra's algorithm with list	$O(V^2)$	Leyzorek et al. 1957, Dijkstra 1959
Dijkstra's algorithm with modified binary heap	$O((E + V) \log V)$	
. . .	. . .	. . .
Dijkstra's algorithm with Fibonacci heap	$O(E + V \log V)$	Fredman & Tarjan 1984, Fredman & Tarjan 1987
	$O(E \log \log L)$	Johnson 1981, Karlsson & Poblete 1983
Gabow's algorithm	$O(E \log_{E/V} L)$	Gabow 1983, Gabow 1985
	$O(E + V \sqrt{\log L})$	Ahuja et al. 1990
Thorup	$O(E + V \log \log V)$	Thorup 2004

Planar Directed Graphs with Arbitrary Weights

All-pairs Shortest Paths

The all-pairs shortest path problem finds the shortest paths between every pair of vertices v, v' in the graph. The all-pairs shortest paths problem for unweighted directed graphs was introduced by Shimbel (1953), who observed that it could be solved by a linear number of matrix multiplications that takes a total time of $O(V^4)$.

Undirected Graph

Weights	Time complexity	Algorithm
R_+	$O(V^3)$	Floyd-Warshall algorithm
N	$O(V^3 / 2^{\Omega(\log n)^{1/2}})$	Williams 2014
R_+	$O(EV \log \alpha(E,V))$	Pettie & Ramachandran 2002
N	$O(EV)$	Thorup 1999 (requires constant-time multiplication).

Directed Graph

Weights	Time complexity	Algorithm
R (no negative cycles)	$O(V^3)$	Floyd-Warshall algorithm
N	$O(V^3 / 2^{\Omega(\log n)^{1/2}})$	Williams 2014
R (no negative cycles)	$O(EV + V^2 \log V)$	Johnson-Dijkstra
R (no negative cycles)	$O(EV + V^2 \log \log V)$	Pettie 2004
N	$O(EV + V^2 \log \log V)$	Hagerup 2000

Applications

Shortest path algorithms are applied to automatically find directions between physical locations, such as driving directions on web mapping websites like MapQuest or Google Maps. For this application fast specialized algorithms are available.

If one represents a nondeterministic abstract machine as a graph where vertices describe states and edges describe possible transitions, shortest path algorithms can be used to find an optimal sequence of choices to reach a certain goal state, or to establish lower bounds on the time needed to reach a given state. For example, if vertices represent the states of a puzzle like a Rubik's Cube and each directed edge corresponds to a single move or turn, shortest path algorithms can be used to find a solution that uses the minimum possible number of moves.

In a networking or telecommunications mindset, this shortest path problem is sometimes called the min-delay path problem and usually tied with a widest path problem. For example, the algorithm may seek the shortest (min-delay) widest path, or widest shortest (min-delay) path.

A more lighthearted application is the games of "six degrees of separation" that try to find the shortest path in graphs like movie stars appearing in the same film.

Other applications, often studied in operations research, include plant and facility layout, robotics, transportation, and VLSI design".

Road Networks

A road network can be considered as a graph with positive weights. The nodes represent road junctions and each edge of the graph is associated with a road segment between two junctions. The weight of an edge may correspond to the length of the associated road segment, the time needed to traverse the segment, or the cost of traversing the segment. Using directed edges it is also possible to model one-way streets. Such graphs are special in the sense that some edges are more important than others for long distance travel (e.g. highways). This property has been formalized using the notion of highway dimension. There are a great number of algorithms that exploit this property and are therefore able to compute the shortest path a lot quicker than would be possible on general graphs.

All of these algorithms work in two phases. In the first phase, the graph is preprocessed without knowing the source or target node. The second phase is the query phase. In this phase, source and target node are known.The idea is that the road network is static, so the preprocessing phase can be done once and used for a large number of queries on the same road network.

The algorithm with the fastest known query time is called hub labeling and is able to compute shortest path on the road networks of Europe or the USA in a fraction of a microsecond. Other techniques that have been used are:

- ALT

- Arc Flags

- Contraction hierarchies

- Transit Node Routing

- Reach based Pruning

- Labeling

Related Problems

The travelling salesman problem is the problem of finding the shortest path that goes through every vertex exactly once, and returns to the start. Unlike the shortest path problem, which can be solved in polynomial time in graphs without negative cycles, the travelling salesman problem is NP-complete and, as such, is believed not to be efficiently solvable for large sets of data. The problem of finding the longest path in a graph is also NP-complete.

The Canadian traveller problem and the stochastic shortest path problem are generalizations where either the graph isn't completely known to the mover, changes over time, or where actions (traversals) are probabilistic.

The shortest multiple disconnected path is a representation of the primitive path network within the framework of Reptation theory.

The widest path problem seeks a path so that the minimum label of any edge is as large as possible.

Strategic Shortest-paths

Sometimes, the edges in a graph have personalities: each edge has its own selfish interest. An example is a communication network, in which each edge is a computer that possibly belongs to a different person. Different computers have different transmission speeds, so every edge in the network has a numeric weight equal to the number of milliseconds it takes to transmit a message. Our goal is to send a message between two points in the network in the shortest time possible. If we know the transmission-time of each computer (-the weight of each edge), then we can use a standard shortest-paths algorithm. If we do not know the transmission times, then we have to ask each computer to tell us its transmission-time. But, the computers may be selfish: a computer might tell us that its transmission time is very long, so that we will not bother it with our messages. A possible solution to this problem is to use a variant of the VCG mechanism, which gives the computers an incentive to reveal their true weights.

Linear Programming Formulation

There is a natural linear programming formulation for the shortest path problem, given below. It is very simple compared to most other uses of linear programs in discrete optimization, however it illustrates connections to other concepts.

Given a directed graph (V, A) with source node s, target node t, and cost w_{ij} for each edge (i, j) in A, consider the program with variables x_{ij}

$$\text{minimize} \sum_{ij \in A} w_{ij} x_{ij} \text{ subject to } x \geq 0 \text{ and for all } i, \sum_j x_{ij} - \sum_j x_{ji} = \begin{cases} 1, & \text{if } i = s; \\ -1, & \text{if } i = t; \\ 0, & \text{otherwise.} \end{cases}$$

The intuition behind this is that x_{ij} is an indicator variable for whether edge (i, j) is part of the shortest path: 1 when it is, and 0 if it is not. We wish to select the set of edges with minimal weight, subject to the constraint that this set forms a path from s to t (represented by the equality constraint: for all vertices except s and t the number of incoming and outcoming edges that are part of the path must be the same (i.e., that it should be a path from s to t).

This LP has the special property that it is integral; more specifically, every basic optimal solution (when one exists) has all variables equal to 0 or 1, and the set of edges whose variables equal 1 form an s-t dipath. Although the origin of this approach dates back to mid-20th century.

The dual for this linear program is

$$\text{maximize } y_t - y_s \text{ subject to for all } ij, y_j - y_i \leq w_{ij}$$

and feasible duals correspond to the concept of a consistent heuristic for the A* algorithm for shortest paths. For any feasible dual y the reduced costs $w'_{ij} = w_{ij} - y_j + y_i$ are nonnegative and A* essentially runs Dijkstra's algorithm on these reduced costs.

General Algebraic Framework on Semirings: The Algebraic Path Problem

Many problems can be framed as a form of the shortest path for some suitably substituted notions of addition along a path and taking the minimum. The general approach to these is to consider the two operations to be those of a semiring. Semiring multiplication is done along the path, and the addition is between paths. This general framework is known as the algebraic path problem.

Most of the classic shortest-path algorithms (and new ones) can be formulated as solving linear systems over such algebraic structures.

More recently, an even more general framework for solving these (and much less obviously related problems) has been developed under the banner of valuation algebras.

Shortest Path in Stochastic Time-dependent Networks

In real-life situations, the transportation network is usually stochastic and time-dependent. In fact, a traveler traversing a link daily may experiences different travel times on that link due not only to the fluctuations in travel demand (origin-destination matrix) but also due to such incidents as work zones, bad weather conditions, accidents and vehicle breakdowns. As a result, a stochastic time-dependent (STD) network is a more realistic representation of an actual road network compared with the deterministic one.

Despite considerable progress during the course of the past decade, it remains a controversial question how an optimal path should be defined and identified in stochastic road networks. In other words, there is no unique definition of an optimal path under uncertainty. One possible and common answer to this question is to find a path with the minimum expected travel time. The main advantage of using this approach is that efficient shortest path algorithms introduced for the deterministic networks can be readily employed to identify the path with the minimum expected travel time in a stochastic network. However, the resulting optimal path identified by this approach may not be reliable, because this approach fails to address travel time variability. It should be noted that the concept of travel time reliability is used interchangeably with travel time variability in the transportation research literature, so that, in general, one can say that the higher the variability in travel time, the lower the reliability would be, and vice versa.

In order to account for travel time reliability more accurately, two common alternative definitions for an optimal path under uncertainty have been suggested. Some have introduced the concept of the most reliable path, aiming to maximize the probability of arriving on time or earlier than a given travel time budget. Others, alternatively, have put forward the concept of an α-reliable path based on which they intended to minimize the travel time budget required to ensure a pre-specified on-time arrival probability.

References

- Paul, Richard (1981). Robot manipulators: mathematics, programming, and control : the computer control of robot manipulators. Cambridge, MA: MIT Press. ISBN 978-0-262-16082-7.

- Spong, Mark W.; Vidyasagar, M. (1989). Robot Dynamics and Control. New York: John Wiley & Sons. ISBN 9780471503521.

- Khalil, Wisama; Dombre, Etienne (2002). Modeling, identification and control of robots. New York: Taylor Francis. ISBN 1-56032-983-1.

- Lipkin, Harvey (2005). "Volume 7: 29th Mechanisms and Robotics Conference, Parts a and B". 2005: 921–926. doi:10.1115/DETC2005-85460. ISBN 0-7918-4744-6.

- Waldron, Kenneth; Schmiedeler, James (2008). "Springer Handbook of Robotics": 9–33. doi:10.1007/978-3-540-30301-5_2. ISBN 978-3-540-23957-4.

- "Computerized Numerical Control". www.sheltonstate.edu. Shelton State Community College. Retrieved March 24, 2015.

- Mike Lynch, "Key CNC Concept #1—The Fundamentals Of CNC", Modern Machine Shop, 4 Jan-uary 1997. Accessed 11 February 2015.

Allied Fields Related to Mechatronics

This chapter is a compilation of the various allied branches of mechatronics that form an integral part of the broader subject matter. The fields described include biomechatronics, cybernetics, ecomechatronics and electromechanics. This content details the practice, technology and methodology of each of these fields.

Biomechatronics

Biomechatronics is an applied interdisciplinary science that aims to integrate biology, mechanics, and electronics. It also encompasses the fields of robotics and neuroscience. Biomechatronic devices encompass a wide range of applications from the development of prosthetic limbs to engineering solutions concerning respiration, vision, and the cardiovascular system.

How it Works

Biomechatronics mimics how the human body works. For example, four different steps must occur to be able to lift the foot to walk. First, impulses from the motor center of the brain are sent to the foot and leg muscles. Next the nerve cells in the feet send information, providing feedback to the brain, enabling it to adjust the muscle groups or amount of force required to walk across the ground. Different amounts of force are applied depending on the type of surface being walked across. The leg's muscle spindle nerve cells then sense and send the position of the floor back up to the brain. Finally, when the foot is raised to step, signals are sent to muscles in the leg and foot to set it down.

Biosensors

Biosensors are used to detect what the user wants to do or their intentions and motions. In some devices the information can be relayed by the user's nervous system or muscle system. This information is related by the biosensor to a controller which can be located inside or outside the biomechatronic device. In addition biosensors receive information about the limb position and force from the limb and actuator. Biosensors come in a variety of forms. They can be wires which detect electrical activity, needle electrodes implanted in muscles, and electrode arrays with nerves growing through them.

Mechanical Sensors

The purpose of the mechanical sensors is to measure information about the biomechatronic device and relate that information to the biosensor or controller.

Controller

The controller in a biomechatronic device relays the user's intentions to the actuators. It also interprets feedback information to the user that comes from the biosensors and mechanical sensors. The other function of the controller is to control the biomechatronic device's movements.

Actuator

The actuator is an artificial muscle. Its job is to produce force and movement. Depending on whether the device is orthotic or prosthetic the actuator can be a motor that assists or replaces the user's original muscle.

Research

Biomechatronics is a rapidly growing field but as of now there are very few labs which conduct research. The Rehabilitation Institute of Chicago, University of California at Berkeley, MIT, and University of Twente in the Netherlands are the researching leaders in biomechatronics. Three main areas are emphasized in the current research.

1. Analyzing human motions, which are complex, to aid in the design of biomechatronic devices

2. Studying how electronic devices can be interfaced with the nervous system.

3. Testing the ways to use living muscle tissue as actuators for electronic devices

Analyzing Motions

A great deal of analysis over human motion is needed because human movement is very complex. MIT and the University of Twente are both working to analyze these movements. They are doing this through a combination of computer models, camera systems, and electromyograms.

Interfacing

Interfacing allows biomechatronic devices to connect with the muscle systems and nerves of the user in order send and receive information from the device. This is a technology that is not available in ordinary orthotics and prosthetics devices. Groups at the University of Twente are making drastic steps in this department. Scientists there have developed a device which will help to treat paralysis and stroke victims who are unable to control their foot while walking. The researchers are also nearing a break-through which would allow a person with an amputated leg to control their prosthetic leg through their stump muscles.

MIT Research

Hugh Herr is the leading biomechatronic scientist at MIT. Herr and his group of re-searchers are developing a sieve integrated circuit electrode and prosthetic devices that are coming closer to mimicking real human movement. The two prosthetic devices cur-rently in the making will control knee movement and the other will control the stiffness of an ankle joint.

Robotic Fish

As mentioned before Herr and his colleagues made a robotic fish that was propelled by living muscle tissue taken from frog legs. The robotic fish was a prototype of a biome-chatronic device with a living actuator. The following characteristics were given to the fish.

- A styrofoam float so the fish can float

- Electrical wires for connections

- A silicone tail that enables force while swimming

- Power provided by lithium batteries

- A microcontroller to control movement

- An infrared sensor enables the microcontroller to communicate with a hand-held device

- Muscles stimulated by an electronic unit

Arts Research

New media artists at UCSD are using biomechatronics in performance art pieces, such as Technesexual (more information, photos, video), a performance which uses bio-metric sensors to bridge the performers' real bodies to their Second Life avatars and Slapshock (more information, photos,video), in which medical TENS units are used to explore intersubjective symbiosis in intimate relationships.

Growth

The demand for biomechatronic devices are at an all-time high and show no signs of slowing down. With increasing technological advancement in recent years, biome-chatronic researchers have been able to construct prosthetic limbs that are capable of replicating the functionality of human appendages. Such devices include the "i-limb", developed by prosthetic company Touch Bionics, the first fully functioning prosthetic hand with articulating joints, as well as Herr's PowerFoot BiOM, the first prosthetic leg capable of simulating muscle and tendon processes within the human body. Biomechatronic research has also helped further research towards understanding human functions. Researchers from Carnegie Mellon and North Carolina State have created an exoskeleton that decreases the metabolic cost of walking by around 7 percent.

Many biomechatronic researchers are closely collaborating with military organizations. The US Department of Veterans Affairs and the Department of Defense are giving funds to different labs to help soldiers and war veterans.

Despite the demand, however, biomechatronic technologies struggle within the healthcare market due to high costs and lack of implementation into insurance policies. Herr claims that Medicare and Medicaid specifically are important "market-breakers or market-makers for all these technologies," and that the technologies will not be available to everyone until the technologies get a breakthrough. Biomechatronic devices, although improved, also still face mechanical obstructions, suffering from inadequate battery power, consistent mechanical reliability, and neural connections between prosthetics and the human body.

Cybernetics

Cybernetics is a transdisciplinary approach for exploring regulatory systems, their structures, constraints, and possibilities. In the 21st century, the term is often used in a rather loose way to imply "control of any system using technology."

Cybernetics is relevant to the study of systems, such as mechanical, physical, biological, cognitive, and social systems. Cybernetics is applicable when a system being analyzed incorporates a closed signaling loop; that is, where action by the system generates some change in its environment and that change is reflected in that system in some manner (feedback) that triggers a system change, originally referred to as a "circular causal" relationship.

System dynamics, a related field, originated with applications of electrical engineering control theory to other kinds of simulation models (especially business systems) by Jay Forrester at MIT in the 1950s.

Concepts studied by cyberneticists include, but are not limited to: learning, cognition, adaptation, social control, emergence, communication, efficiency, efficacy, and connectivity. These concepts are studied by other subjects such as engineering and biology, but in cybernetics these are abstracted from the context of the individual organism or device.

Norbert Wiener defined cybernetics in 1948 as "the scientific study of control and communication in the animal and the machine." Contemporary cybernetics began as an interdisciplinary study connecting the fields of control systems, electrical network theory, mechanical engineering, logic modeling, evolutionary biology, neuroscience, anthropology, and psychology in the 1940s, often attributed to the Macy Conferences. During the second half of the 20th century cybernetics evolved in ways that distinguish first-order cybernetics (about observed systems) from second-order cybernetics (about observing systems). More recently there is talk about a third-order cybernetics (doing in ways that embraces first and second-order).

Fields of study which have influenced or been influenced by cybernetics include game theory, system theory (a mathematical counterpart to cybernetics), perceptual control theory, sociology, psychology (especially neuropsychology, behavioral psychology, cognitive psychology), philosophy, architecture, and organizational theory.

Definitions

Cybernetics has been defined in a variety of ways, by a variety of people, from a variety of disciplines. The *Larry Richards Reader* includes a listing by Stuart Umpleby of notable definitions:

- "Science concerned with the study of systems of any nature which are capable of receiving, storing and processing information so as to use it for control." — A. N. Kolmogorov

- "The art of securing efficient operation." — Louis Couffignal

- "'The art of steersmanship': deals with all forms of behavior in so far as they are regular, or determinate, or reproducible: stands to the real machine -- electronic, mechanical, neural, or economic -- much as geometry stands to real object in our terrestrial space; offers a method for the scientific treatment of the system in which complexity is outstanding and too important to be ignored." — W. Ross Ashby

- "A branch of mathematics dealing with problems of control, recursiveness, and information, focuses on forms and the patterns that connect." — Gregory Bateson

- "The art of effective organization." — Stafford Beer

- "The art and science of manipulating defensible metaphors." — Gordon Pask

- "The art of creating equilibrium in a world of constraints and possibilities." — Ernst von Glasersfeld

- "The science and art of understanding." — Humberto Maturana

- "The ability to cure all temporary truth of eternal triteness." — Herbert Brun

Other notable definitions include:

- "The science and art of the understanding of understanding." — Rodney E. Donaldson, the first president of the American Society for Cybernetics

- "The control of an automaton's feedback loop." - Link Starbureiy

- "A way of thinking about ways of thinking of which it is one." — Larry Richards

- "The art of interaction in dynamic networks." — Roy Ascott

Etymology

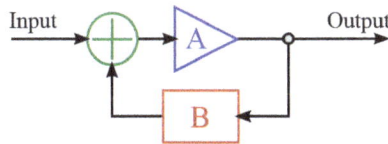

Simple feedback model. AB < 0 for negative feedback.

As with the ancient Greek pilot, independence of thought is important in cybernetics. Cybernetics is a broad field of study, but the essential goal of cybernetics is to understand and define the functions and processes of systems that have goals and that participate in circular, causal chains that move from action to sensing to comparison with desired goal, and again to action. Studies in cybernetics provide a means for examining the design and function of any system, including social systems such as business management and organizational learning, including for the purpose of making them more efficient and effective.

French physicist and mathematician André-Marie Ampère first coined the word "cybernetique" in his 1834 essay *Essai sur la philosophie des sciences* to describe the science of civil government.

Cybernetics was borrowed by Norbert Wiener, in his book *Cybernetics*, to define the study of control and communication in the animal and the machine. Stafford Beer called it the science of effective organization and Gordon Pask called it "the art of defensible metaphors" (emphasizing its constructivist epistemology) though he later extended it to include information flows "in all media" from stars to brains. It includes the study of feedback, black boxes and derived concepts such as communication and control in living organisms, machines and organizations including self-organization. Its focus is how anything (digital, mechanical or biological) processes information, reacts to information, and changes or can be changed to better accomplish the first two tasks. A more philosophical definition, suggested in 1956 by Louis Couffignal, one of the pioneers of cybernetics, characterizes cybernetics as "the art of ensuring the efficacy of action." The most recent definition has been proposed by Louis Kauffman, President of the American Society for Cybernetics, "Cybernetics is the study of systems and processes that interact with themselves and produce themselves from themselves."

History

Roots of Cybernetic Theory

The word *cybernetics* was first used in the context of "the study of self-governance" by Plato in The Alcibiades to signify the governance of people. The word 'cybernétique' was also used in 1834 by the physicist André-Marie Ampère (1775–1836) to denote the sciences of government in his classification system of human knowledge.

James Watt

The first artificial automatic regulatory system, a water clock, was invented by the mechanician Ktesibios. In his water clocks, water flowed from a source such as a holding tank into a reservoir, then from the reservoir to the mechanisms of the clock. Ktesibios's device used a cone-shaped float to monitor the level of the water in its reservoir and adjust the rate of flow of the water accordingly to maintain a constant level of water in the reservoir, so that it neither overflowed nor was allowed to run dry. This was the first artificial truly automatic self-regulatory device that required no outside intervention between the feedback and the controls of the mechanism. Although they did not refer to this concept by the name of Cybernetics (they considered it a field of engineering), Ktesibios and others such as Heron and Su Song are considered to be some of the first to study cybernetic principles.

The study of *teleological mechanisms* in machines with *corrective feedback* dates from as far back as the late 18th century when James Watt's steam engine was equipped with a governor (1775-1800), a centrifugal feedback valve for controlling the speed of the engine. Alfred Russel Wallace identified this as the principle of evolution in his famous 1858 paper. In 1868 James Clerk Maxwell published a theoretical article on governors, one of the first to discuss and refine the principles of self-regulating devices. Jakob von Uexküll applied the feedback mechanism via his model of functional cycle (*Funktionskreis*) in order to explain animal behaviour and the origins of meaning in general.

Early 20th Century

Contemporary cybernetics began as an interdisciplinary study connecting the fields of control systems, electrical network theory, mechanical engineering, logic modeling, evolutionary biology and neuroscience in the 1940s. Electronic control systems originated with the 1927 work of Bell Telephone Laboratories engineer Harold S. Black on using negative feedback to control amplifiers. The ideas are also related to the biological work of Ludwig von Bertalanffy in General Systems Theory.

Early applications of negative feedback in electronic circuits included the control of gun mounts and radar antenna during World War II. Jay Forrester, a graduate student at the Servomechanisms Laboratory at MIT during WWII working with Gordon S. Brown to develop electronic control systems for the U.S. Navy, later applied these ideas to social organizations such as corporations and cities as an original organizer of the MIT School of Industrial Management at the MIT Sloan School of Management. Forrester is known as the founder of System Dynamics.

W. Edwards Deming, the Total Quality Management guru for whom Japan named its top post-WWII industrial prize, was an intern at Bell Telephone Labs in 1927 and may have been influenced by network theory. Deming made "Understanding Systems" one of the four pillars of what he described as "Profound Knowledge" in his book "The New Economics."

Numerous papers spearheaded the coalescing of the field. In 1935 Russian physiologist P.K. Anokhin published a book in which the concept of feedback ("back afferentation") was studied. The study and mathematical modelling of regulatory processes became a continuing research effort and two key articles were published in 1943. These papers were "Behavior, Purpose and Teleology" by Arturo Rosenblueth, Norbert Wiener, and Julian Bigelow; and the paper "A Logical Calculus of the Ideas Immanent in Nervous Activity" by Warren McCulloch and Walter Pitts.

In 1936, Odobleja publishes "Phonoscopy and the clinical semiotics". In 1937, he participates in the IXth International Congress of Military Medicine with a paper entitled "Demonstration de phonoscopie", where he disseminates a prospectus in French, announcing the appearance of his future work "The Consonantist Psychology".

The most important of his writings is Psychologie consonantiste, in which Odobleja lays the theoretical foundations of the generalized cybernetics. The book, published in Paris by "Librairie Maloine" (vol. I in 1938 and vol. II in 1939), contains almost 900 pages and includes 300 figures in the text. The author wrote at the time that "this book is... a table of contents, an index or a dictionary of psychology, [for] a ... great Treatise of Psychology that should contain 20–30 volumes".

Due to the beginning of World War II, the publication went unnoticed. The first Romanian edition of this work did not appear until 1982 (the first edition was published in French). Cybernetics as a discipline was firmly established by Norbert Wiener, McCulloch and others, such as W. Ross Ashby, mathematician Alan Turing, and W. Grey Walter. Walter was one of the first to build autonomous robots as an aid to the study of animal behaviour. Together with the US and UK, an important geographical locus of early cybernetics was France.

In the spring of 1947, Wiener was invited to a congress on harmonic analysis, held in Nancy, France. The event was organized by the Bourbaki, a French scientific society, and mathematician Szolem Mandelbrojt (1899–1983), uncle of the world-famous mathematician Benoît Mandelbrot.

John von Neumann

During this stay in France, Wiener received the offer to write a manuscript on the unifying character of this part of applied mathematics, which is found in the study of Brownian motion and in telecommunication engineering. The following summer, back in the United States, Wiener decided to introduce the neologism cybernetics into his scientific theory. The name *cybernetics* was coined to denote the study of "teleological mechanisms" and was popularized through his book *Cybernetics, or Control and Communication in the Animal and the Machine* (MIT Press/John Wiley and Sons, NY, 1948). In the UK this became the focus for the Ratio Club.

In the early 1940s John von Neumann, although better known for his work in mathematics and computer science, did contribute a unique and unusual addition to the world of cybernetics: von Neumann cellular automata, and their logical follow up the von Neumann Universal Constructor. The result of these deceptively simple thought-experiments was the concept of self replication which cybernetics adopted as a core concept. The concept that the same properties of genetic reproduction applied to social memes, living cells, and even computer viruses is further proof of the somewhat surprising universality of cybernetic study.

Wiener popularized the social implications of cybernetics, drawing analogies between automatic systems (such as a regulated steam engine) and human institutions in his best-selling *The Human Use of Human Beings : Cybernetics and Society* (Houghton-Mifflin, 1950).

While not the only instance of a research organization focused on cybernetics, the Biological Computer Lab at the University of Illinois, Urbana/Champaign, under the direction of Heinz von Foerster, was a major center of cybernetic research for almost 20 years, beginning in 1958.

Split from Artificial Intelligence

Artificial intelligence (AI) was founded as a distinct discipline at the Dartmouth Conferences. After some uneasy coexistence, AI gained funding and prominence. Consequently, cybernetic sciences such as the study of neural networks were downplayed; the discipline shifted into the world of social sciences and therapy.

Prominent cyberneticians during this period include:

- Gregory Bateson

- Aksel Berg

New Cybernetics

In the 1970s, new cyberneticians emerged in multiple fields, but especially in biology. The ideas of Maturana, Varela and Atlan, according to Jean-Pierre Dupuy (1986) "realized that the cybernetic metaphors of the program upon which molecular biology had been based rendered a conception of the autonomy of the living being impossible. Consequently, these thinkers were led to invent a new cybernetics, one more suited to the organizations which mankind discovers in nature - organizations he has not himself invented". However, during the 1980s the question of whether the features of this new cybernetics could be applied to social forms of organization remained open to debate.

In political science, Project Cybersyn attempted to introduce a cybernetically controlled economy during the early 1970s. In the 1980s, according to Harries-Jones (1988) "unlike its predecessor, the new cybernetics concerns itself with the interaction of autonomous political actors and subgroups, and the practical and reflexive consciousness of the subjects who produce and reproduce the structure of a political community. A dominant consideration is that of recursiveness, or self-reference of political action both with regards to the expression of political consciousness and with the ways in which systems build upon themselves".

One characteristic of the emerging new cybernetics considered in that time by Felix Geyer and Hans van der Zouwen, according to Bailey (1994), was "that it views information as constructed and reconstructed by an individual interacting with the environment. This provides an epistemological foundation of science, by viewing it as observer-dependent. Another characteristic of the new cybernetics is its contribution towards bridging the *micro-macro gap*. That is, it links the individual with the society". Another characteristic noted was the "transition from classical cybernetics to the new cybernetics [that] involves a transition from classical problems to new problems. These shifts in thinking involve, among others, (a) a change from emphasis on the system being steered to the system doing the steering, and the factor which guides the steering decisions.; and (b) new emphasis on communication between several systems which are trying to steer each other".

Recent endeavors into the true focus of cybernetics, systems of control and emergent behavior, by such related fields as game theory (the analysis of group interaction), systems of feedback in evolution, and metamaterials (the study of materials with properties beyond the Newtonian properties of their constituent atoms), have led to a revived interest in this increasingly relevant field.

Cybernetics and Economic Systems

The design of self-regulating control systems for a real-time planned economy was explored by Viktor Glushkov in the former Soviet Union during the 1960s. By the time information technology was developed enough to enable feasible economic planning based on computers, the Soviet Union and eastern bloc countries began moving away from planning and eventually collapsed.

More recent proposals for socialism involve "New Socialism", outlined by the computer scientists Paul Cockshott and Allin Cottrell, where computers determine and manage the flows and allocation of resources among socially-owned enterprises.

Subdivisions of the Field

Cybernetics is sometimes used as a generic term, which serves as an umbrella for many systems-related scientific fields.

Basic Cybernetics

Cybernetics studies systems of control as a concept, attempting to discover the basic principles underlying such things as

ASIMO uses sensors and sophisticated algorithms to avoid obstacles and navigate stairs.

- Artificial intelligence

- Computer vision

- Control systems

- Conversation theory

- Emergence

- Interactions of actors theory

- Learning organization

- Robotics

- Second-order cybernetics

- Self-organization in cybernetics

In Biology

Cybernetics in biology is the study of cybernetic systems present in biological organisms, primarily focusing on how animals adapt to their environment, and how information in the form of genes is passed from generation to generation. There is also a secondary focus on combining artificial systems with biological systems. A notable application to the biology world would be that, in 1955, the physicist George Gamow published a prescient article in *Scientific American* called "Information transfer in the living cell", and cybernetics gave biologists Jacques Monod and François Jacob a language for formulating their early theory of gene regulatory networks in the 1960s.

- Autopoiesis

- Biocybernetics

- Bioengineering

- Bionics

- Heterostasis

- Homeostasis

- Medical cybernetics

- Neuroscience

- Practopoiesis

- Synthetic biology

- Systems biology

In computer Science

Computer science directly applies the concepts of cybernetics to the control of devices and the analysis of information.

- Cellular automaton

- Decision support system

- Design patterns

- Robotics

- Simulation

In Engineering

Cybernetics in engineering is used to analyze cascading failures and system accidents, in which the small errors and imperfections in a system can generate disasters. Other topics studied include:

An artificial heart, a product of biomedical engineering.

- Adaptive systems

- Biomedical engineering

- Engineering cybernetics

- Ergonomics

- Systems engineering

In Management

- Entrepreneurial cybernetics

- Management cybernetics

- Operations research

- Organizational cybernetics

- Systems engineering

In Mathematics

Mathematical Cybernetics focuses on the factors of information, interaction of parts in systems, and the structure of systems.

- Control theory

- Dynamical system

- Information theory

- Systems theory

In Psychology

- Attachment theory

- Behavioral cybernetics

- Cognitive psychology

- Consciousness

- Embodied cognition

- Homunculus

- Human-robot interaction

- Mind-body problem

- Perceptual control theory

- Psycho-Cybernetics

- Psychovector analysis

- Systems psychology

In Sociology

By examining group behavior through the lens of cybernetics, sociologists can seek the reasons for such spontaneous events as smart mobs and riots, as well as how communities develop rules such as etiquette by consensus without formal discussion. Affect Control Theory explains role behavior, emotions, and labeling theory in terms of homeostatic maintenance of sentiments associated with cultural categories. The most comprehensive attempt ever made in the social sciences to increase cybernetics in a generalized theory of society was made by Talcott Parsons. In this way, cybernetics establishes the basic hierarchy in Parsons' AGIL paradigm, which is the ordering system-dimension of his action theory. These and other cybernetic models in sociology are reviewed in a book edited by McClelland and Fararo.

- Affect control theory

- Memetics

- Sociocybernetics

In Education

A model of cybernetics in Education was introduced by Gihan Sami Soliman; an educational consultant, as a project idea to be implemented with the help of two team members in Sinai. The Sinai Sustainability Cybernetics Center announced as a semi-finalist project by MIT annual competition 2013. The project idea proposed relating education to sustainable development through an IMS project that applies a multiple educational program related to the original natural self-healing system of life on earth. Education, sustainable development, social justice disciplines interact in a causal circular relationship that education would contribute to the development of the local community in Sinai village, on both sustainability and social responsibility levels while the community itself provides a unique learning environment that will contribute to the development of the educational program in a closed signaling loop.

In Art

Nicolas Schöffer's *CYSP I* (1956) was perhaps the first artwork to explicitly employ cybernetic principles (CYSP is an acronym that joins the first two letters of the words "CYbernetic" and "SPatiodynamic"). The artist Roy Ascott elaborated an extensive theory of cybernetic art in "Behaviourist Art and the Cybernetic Vision" (Cybernetica, Journal of the International Association for Cybernetics (Namur), Volume IX, No.4, 1966; Volume X No.1, 1967) and in "The Cybernetic Stance: My Process and Purpose" (Leonardo Vol 1, No 2, 1968). Art historian Edward A. Shanken has written about the history of art and cybernetics in essays including "Cybernetics and Art: Cultural Convergence in the 1960s" and "From Cybernetics to Telematics: The Art, Pedagogy, and Theory of Roy Ascott"(2003), which traces the trajectory of Ascott's work from cybernetic art to telematic art (art using computer networking as its medium, a precursor to net.art.)

- Telematic art

- Interactive art

- Systems art

In Earth System Science

Geocybernetics aims to study and control the complex co-evolution of ecosphere and anthroposphere, for example, for dealing with planetary problems such as anthropogenic global warming. Geocybernetics applies a dynamical systems perspective to Earth system analysis. It provides a theoretical framework for studying the implications of following different sustainability paradigms on co-evolutionary trajectories of the planetary socio-ecological system to reveal attractors in this system, their stability, resilience and reachability. Concepts such as tipping points in the climate system, planetary boundaries, the safe operating space for humanity and proposals for manipulating Earth system dynamics on a global scale such as geoengineering have been framed in the language of geocybernetic Earth system analysis.

Related Fields

Complexity Science

Complexity science attempts to understand the nature of complex systems.

- Complex adaptive system

- Complex systems

- Complexity theory

Biomechatronics

Biomechatronics relates to linking mechatronics to biological organisms, leading to systems that conform to A. N. Kolmogorov's definition of Cybernetics, i.e. "Science concerned with the study of systems of any nature which are capable of receiving, storing and processing information so as to use it for control". From this perspective mechatronics are considered technical cybernetics or engineering cybernetics.

Ecomechatronics

Ecomechatronics is an engineering approach to developing and applying mechatronical technology in order to reduce the ecological impact and total cost of ownership of machines. It builds upon the integrative approach of mechatronics, but not with the aim of only improving the functionality of a machine. Mechatronics is the multidisciplinary field of science and engineering that merges mechanics, electronics, control theory, and computer science to improve and optimize product design and manufacturing. In ecomechatronics, additionally, functionality should go hand in hand with an efficient use and limited impact on resources. Machine improvements are targeted in 3 key areas: energy efficiency, performance and user comfort (noise & vibrations).

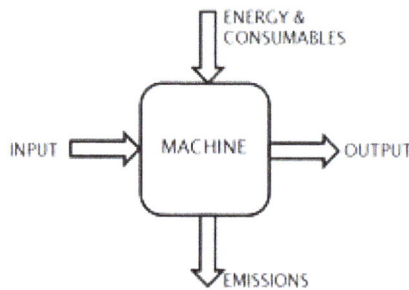

Machine as a system requiring energy and consumables to transform an input into an output, thereby generating emissions (heat, noise, ...)

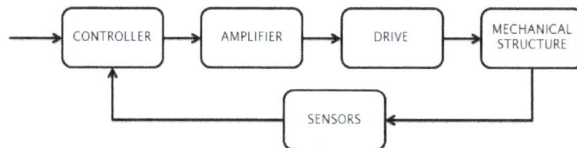

Schematic of a mechatronical system consisting of a controller, amplifier, drive, mechanical structure and sensors

Description

Among policy makers and manufacturing industries there is a growing awareness of the scarcity of resources and the need for sustainable development. This results in new regulations with respect to the design of machines (e.g. European Ecodesign Directive 2009/125/EC) and to a paradigm shift in the global machines market: "instead of maximum profit from minimum capital, maximum added value must be generated from minimal resources". Manufacturing industries increasingly require high performance machines that use resources (energy, consumables) economically in a human-centered production. Machine building companies and original equipment manufacturers are thus urged to respond to this market demand with a new generation of high performance machines with higher energy efficiency and user comfort.

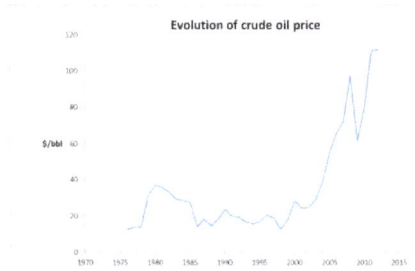

Evolution of crude oil price. Data source: Statistical Review of World Energy 2013, BP

A reduction of the energy consumption lowers energy costs and reduces environmental impact. Typically more than 80% of the total-life-cycle impact of a machine is attributed to its energy consumption during the use phase. Therefore, improving a machine's energy efficiency is the most effective way of reducing its environmental impact. Performance quantifies how well a machine executes its function and is typically related to productivity, precision and availability. User comfort is related to the exposure of operators and the environment to noise & vibrations due to machine operation.

Since energy efficiency, performance and noise & vibrations are coupled in a machine they need to be addressed in an integrated way in the design phase. Example of the interrelation between the 3 key areas: with increasing machine speed typically the machine's productivity increases, but energy consumption will increase as well and machine vibrations may grow such that machine accuracy (e.g. positioning accuracy) and availability (due to downtime and maintenance) decrease. Ecomechatronical design deals with the trade-off between these key areas.

Approach

Ecomechatronics impacts the way mechatronical systems and machines are being designed and implemented. Therefore, the transformation to a new generation of machines concerns knowledge institutes, original equipment manufacturers, CAE software suppliers, machine builders and industrial machine owners. The fact that about 80% of the environmental impact of a machine is determined by its design puts emphasis on making the right technological design choices. A model-based, multidisciplinary design approach is required in order to address the energy efficiency, performance and user comfort of a machine in an integrated way.

The key enabling technologies can be categorized in machine components, machine design methods & tools, and machine control. A few examples are listed below per category.

Machine Components

- Energy efficient electrical motors: cf. energy efficiency classes of electric motors, ecodesign requirements for electric motors

- Variable frequency drives: variable motor speed enables energy reduction with respect to fixed speed applications

- Variable hydraulic pumps: energy reduction by adapting to required pressure and flow (e.g. variable displacement pump, load sensing pump)

- Energy storage technologies: electrical (battery, capacitor, supercapacitor), hydraulical (accumulator), kinetic energy (flywheel), pneumatic, magnetic (superconducting magnetic energy storage)

Design Methods & Tools

- Energetic simulations: using energetic machine models and empirical data (e.g. energy efficiency maps) to estimate the machine's energy consumption in the design phase

- Energy demand optimization: e.g. load leveling in order to avoid peaks in power demand

- Hybridization: applying at least one other, intermediate energy form in order to reduce primary power source consumption e.g. in vehicles with internal combustion engines

- Vibro-acoustic analysis: study of the noise & vibrations signature of a machine in order to localize and differentiate between their root causes

- Multibody modeling: simulation of the interaction forces and displacements of coupled rigid bodies, e.g. to assess the effect of vibration dampers on a mechanical structure

- Active vibration damping: e.g. use of piezoelectric bearings for active control of machine vibrations

- Rapid control prototyping: provides a fast and inexpensive way for control and signal processing engineers to verify designs early and evaluate design tradeoffs

Machine Control

- Energy consumption minimization: control signals are optimized for minimum energy consumption

- Energy management of energy storage systems: controlling the power flows and state-of-charge of an energy storage system with the aim of achieving maximum energy benefit and maximum system lifespan

- Model-based control: taking advantage of system models to improve the outcome (accuracy, reaction time, ...) of the controlled system

- (Self-)learning control: control self-adapting to the system and its changing environment, reducing the need for control parameter tuning and adaptation by the control engineer

- Optimal machine control: the control of the system is regarded as an optimization problem to which the control rules are considered the optimal solution

Applications

Some examples of ecomechatronical system applications are:

- Komatsu PC200-8 Hybrid: the world's first hybrid excavator has an energy storage system based on supercapacitors. The energy recuperation in the hydraulic drive line during braking results in a significant improvement of fuel economy.

- Hybrid bus: different hybrid bus types have been commercialized (e.g. Exqui-City bus by Van Hool), using fuel cells or a diesel engine as a primary energy source and batteries and/or supercapacitors as energy storage systems.

- Hybrid tram vehicle: hybridization in tram vehicles enables energy recuperation as well as mobility without overhead lines, as applied in e.g. some of the Combino Supra tram vehicles by Siemens Transportation Systems. The system uses a combination of traction batteries and supercapacitors.

Electromechanics

In engineering, electromechanics combines electrical and mechanical processes and procedures drawn from electrical engineering and mechanical engineering. Electrical engineering in this context also encompasses electronics engineering.

Devices which carry out electrical operations by using moving parts are known as electromechanical. Strictly speaking, a manually operated switch is an electromechanical component, but the term is usually understood to refer to devices which involve an electrical signal to create mechanical movement, or mechanical movement to create an electric signal. Often involving electromagnetic principles such as in relays, which allow a voltage or current to control other, usually isolated circuit voltage or current by mechanically switching sets of contacts, and solenoids, by which a voltage can actuate a moving linkage as in solenoid valves. Piezoelectric devices are electromechanical, but do not use electromagnetic principles. Piezoelectric devices can create sound or vibration from an electrical signal or create an electrical signal from sound or mechanical vibration.

Before the development of modern electronics, electromechanical devices were widely used in complicated systems subsystems, including electric typewriters, teleprinters, very early television systems, and the very early electromechanical digital computers.

History of Electromechanics

Relays originated with telegraphy as electromechanical devices used to regenerate telegraph signals. In 1885, Michael Pupin at Columbia University taught mathematical physics and electromechanics until 1931.

The Strowger switch, the Panel switch, and similar ones were widely used in early automated telephone exchanges. Crossbar switches were first widely installed in the middle 20th century in Sweden, the United States, Canada, and Great Britain, and these quickly spread to the rest of the world - especially to Japan. The electromechanical television systems of the late 19th century were less successful.

Electric typewriters developed, up to the 1980s, as "power-assisted typewriters". They contained a single electrical component, the motor. Where the keystroke had previously moved a typebar directly, now it engaged mechanical linkages that directed mechanical power from the motor into the typebar. This was also true of the later IBM Selectric. At Bell Labs, in the 1940s, the Bell Model V computer was developed. It was an electromechanical relay-based device; cycles took seconds. In 1968 electromechanical systems were still under serious consideration for an aircraft flight control computer, until a device based on large scale integration electronics was adopted in the Central Air Data Computer.

Modern Practice

Beginning in the last third of the century, much equipment which for most of the 20th century would have used electromechanical devices for control, has come to use less expensive and more reliable integrated microcontroller circuits containing ultimately a few million transistors, and a program to carry out the same task through logic, with electromechanical components only where moving parts, such as mechanical electric actuators, are a requirement. Such chips have replaced most electromechanical devices, because any point in a system which must rely on mechanical movement for proper operation will have mechanical wear and eventually fail. Properly designed electronic circuits without moving parts will continue to operate properly almost indefinitely and are used in most simple feedback control systems, and appear in huge numbers in everything from traffic lights to washing machines.

As of 2010, approximately 16,400 people work as electro-mechanical technicians in the US, about 1 out of every 9000 workers. Their median annual wage is about 50% more than the median annual wage over all occupations.

References

- Ravindra K. Ahuja, Thomas L. Magnanti, and James B. Orlin (1993). Network Flows: Theory, Algorithms and Applications. Prentice Hall. ISBN 0-13-617549-X.

- John Baras; George Theodorakopoulos (4 April 2010). Path Problems in Networks. Morgan & Claypool Publishers. pp. 9–. ISBN 978-1-59829-924-3.

- Michel Gondran; Michel Minoux (2008). Graphs, Dioids and Semirings: New Models and Algorithms. Springer Science & Business Media. chapter 4. ISBN 978-0-387-75450-5.

- Marc Pouly; Jürg Kohlas (2011). Generic Inference: A Unifying Theory for Automated Reasoning. John Wiley & Sons. Chapter 6. Valuation Algebras for Path Problems. ISBN 978-1-118-01086-0.

- Fakoor, Mahdi; Kosari, Amirreza; Jafarzadeh, Mohsen (2016). "Humanoid robot path planning with fuzzy Markov decision processes". Journal of Applied Research and Technology. doi:10.1016/j.jart.2016.06.006. Retrieved 21 August 2016.

- Fanning, Paul (March 13, 2014). "How biomechatronic prosthetics are changing the face of disability". Eureka Magazine. Retrieved July 29, 2016.

- "SSCC (Sinai Sustainability Cybernetics Center)" the 46th team to qualify for this year's MIT semi-finalist round — Naharnet. Naharnet.com (2013-04-25). Retrieved on 2013-11-02.

- Kenny, Vincent (15 March 2009). "There's Nothing Like the Real Thing". Revisiting the Need for a Third-Order Cybernetics". Constructivist Foundations. 4 (2): 100–111. Retrieved 6 June 2012.

- Allin Cottrell & W.Paul Cockshott, Towards a new socialism (Nottingham, England: Spokesman, 1993). Retrieved: 17 March 2012.

Applications of Mechatronics

The encompassing nature and flexibility of mechatronics has ensured its usage in applications that span disciplines. This chapter describes topics like machine vision, expert system, cyber-physical system, anti-lock breaking system, servomechanism, control system, cruise control etc. It explores the methods, uses and operation of these applications.

Machine Vision

Machine vision (MV) is the technology and methods used to provide imaging-based automatic inspection and analysis for such applications as automatic inspection, process control, and robot guidance in industry. The scope of MV is broad. MV is related to, though distinct from, computer vision.

Early Automatix (now part of Microscan) machine vision system Autovision II from 1983 being demonstrated at a trade show. Camera on tripod is pointing down at a light table to produce backlit image shown on screen, which is then subjected to blob extraction.

Applications

The primary uses for machine vision are automatic inspection and industrial robot guidance. Other machine vision applications include:

- Automated Train Examiner (ATEx) Systems
- Automatic PCB inspection

- Wood quality inspection

- Final inspection of sub-assemblies

- Engine part inspection

- Label inspection on products

- Checking medical devices for defects

- Final inspection cells

- Robot guidance and checking orientation of components

- Packaging Inspection

- Medical vial inspection

- Food pack checks

- Verifying engineered components

- Wafer Dicing

- Reading of Serial Numbers

- Inspection of Saw Blades

- Inspection of Ball Grid Arrays (BGAs)

- Surface Inspection

- Measuring of Spark Plugs

- Molding Flash Detection

- Inspection of Punched Sheets

- 3D Plane Reconstruction with Stereo

- Pose Verification of Resistors

- Classification of Non-Woven Fabrics

Methods

Machine vision methods are defined as both the process of defining and creating an MV solution, and as the technical process that occurs during the operation of the solution. Here the latter is addressed. As of 2006, there was little standardization in the interfacing and configurations used in MV. This includes user interfaces, interfaces for the integration of multi-component systems and automated data interchange. Nonetheless, the first step in the MV sequence of operation is acquisition of an image, typically using cameras, lenses, and lighting that has been designed to provide the differentiation required by subsequent processing. MV software packages then employ various digital image processing techniques to extract the required information, and often make decisions (such as pass/fail) based on the extracted information.

Imaging

While conventional (2D visible light) imaging is most commonly used in MV, alternatives include imaging various infrared bands, line scan imaging, 3D imaging of surfaces and X-ray imaging. Key divisions within MV 2D visible light imaging are monochromatic vs. color, resolution, and whether or not the imaging process is simultaneous over the entire image, making it suitable for moving processes. The most commonly used method for 3D imaging is scanning based triangulation which utilizes motion of the product or image during the imaging process. Other 3D methods used for machine vision are time of flight, grid based and stereoscopic.

The imaging device (e.g. camera) can either be separate from the main image processing unit or combined with it in which case the combination is generally called a smart camera or smart sensor. When separated, the connection may be made to specialized intermediate hardware, a frame grabber using either a standardized (Camera Link, CoaXPress) or custom interface. MV implementations also have used digital cameras capable of direct connections (without a framegrabber) to a computer via FireWire, USB or Gigabit Ethernet interfaces.

Though the vast majority of machine vision applications are solved using two-dimensional imaging, machine vision applications utilizing 3D imaging are a growing niche within the industry. One method is grid array based systems using pseudorandom structured light system as employed by the Microsoft Kinect system circa 2012. Another method of generating a 3D image is to use laser triangulation, where a laser is projected onto the surfaces of an object and the deviation of the line is used to calculate the shape. In machine vision this is accomplished with a scanning motion, either by moving the workpiece, or by moving the camera & laser imaging system. Stereoscopic vision is used in special cases involving unique features present in both views of a pair of cameras.

Image Processing

After an image is acquired, it is processed. Machine vision image processing methods include

- Stitching/Registration: Combining of adjacent 2D or 3D images.

- Filtering (e.g. morphological filtering)

- Thresholding: Thresholding starts with setting or determining a gray value that will be useful for the following steps. The value is then used to separate portions of the image, and sometimes to transform each portion of the image simply black and white based on whether it is below or above that grayscale value.

- Pixel counting: counts the number of light or dark pixels

- Segmentation: Partitioning a digital image into multiple segments to simplify and/or change the representation of an image into something that is more meaningful and easier to analyze.

- Inpainting

- Edge detection: finding object edges

- Color Analysis: Identify parts, products and items using color, assess quality from color, and isolate features using color.

- Blob discovery & manipulation: inspecting an image for discrete blobs of connected pixels (e.g. a black hole in a grey object) as image landmarks. These blobs frequently represent optical targets for machining, robotic capture, or manufacturing failure.

- Neural net processing: weighted and self-training multi-variable decision making

- Pattern recognition including template matching. Finding, matching, and/or counting specific patterns. This may include location of an object that may be rotated, partially hidden by another object, or varying in size.

- Barcode, Data Matrix and "2D barcode" reading

- Optical character recognition: automated reading of text such as serial numbers

- Gauging/Metrology: measurement of object dimensions (e.g. in pixels, inches or millimeters)

- Comparison against target values to determine a "pass or fail" or "go/no go" result. For example, with code or bar code verification, the read value is compared to the stored target value. For gauging, a measurement is compared against the proper value and tolerances. For verification of alpha-numberic codes, the OCR'd value is compared to the proper or target value. For inspection for blemishes, the measured size of the blemishes may be compared to the maximums allowed by quality standards.

Outputs

A common output from machine vision systems is pass/fail decisions. These decisions may in turn trigger mechanisms that reject failed items or sound an alarm. Other common outputs include object position and orientation information from robot guidance systems. Additionally, output types include numerical measurement data, data read from codes and characters, displays of the process or results, stored images, alarms from automated space monitoring MV systems, and process control signals.

Market

As recently as 2006, one industry consultant reported that MV represented a $1.5 billion market in North America. However, the editor-in-chief of an MV trade magazine asserted that "machine vision is not an industry per se" but rather "the integration of technologies and products that provide services or applications that benefit true industries such as automotive or consumer goods manufacturing, agriculture, and defense."

As of 2006, experts estimated that MV had been employed in less than 20% of the applications for which it is potentially useful.

For the latest figures for the European sector: http://www.emva.org/vision-insights/market-data/quarterly-european-vision-sales-report/

Expert System

A Symbolics Lisp Machine: An Early Platform for Expert Systems.
Note the unusual "space cadet keyboard".

In artificial intelligence, an expert system is a computer system that emulates the decision-making ability of a human expert. Expert systems are designed to solve complex problems by reasoning about knowledge, represented primarily as if–then rules rather than through conventional procedural code. The first expert systems were created in the 1970s and then proliferated in the 1980s. Expert systems were among the first truly successful forms of AI software.

An expert system is divided into two sub-systems: the inference engine and the knowledge base. The knowledge base represents facts and rules. The inference engine applies the rules to the known facts to deduce new facts. Inference engines can also include explanation and debugging capabilities.

History

Edward Feigenbaum said that the key insight of early expert systems was that "intelligent systems derive their power from the knowledge they possess rather than from the specific formalisms and inference schemes they use." Although, in retrospect, this seems a rather straightforward insight, it was a significant step forward at the time. Until then, research had been focused on attempts to develop very general-purpose problem solvers such as those described by Allen Newell and Herb Simon.

Expert systems were introduced by the Stanford Heuristic Programming Project led by Feigenbaum, who is sometimes referred to as the "father of expert systems". The Stanford researchers tried to identify domains where expertise was highly valued and complex, such as diagnosing infectious diseases (Mycin) and identifying unknown organic molecules (Dendral).

In addition to Feigenbaum key early contributors were Edward Shortliffe, Bruce Buchanan, and Randall Davis. Expert systems were among the first truly successful forms of AI software.

Research on expert systems was also active in France. In the US the focus tended to be on rule-based systems, first on systems hard coded on top of LISP programming environments and then on expert system shells developed by vendors such as Intellicorp. In France research focused more on systems developed in Prolog. The advantage of expert system shells was that they were somewhat easier for non-programmers to use. The advantage of Prolog environments was that they weren't focused only on IF-THEN rules. Prolog environments provided a much fuller realization of a complete First Order Logic environment.

In the 1980s, expert systems proliferated. Universities offered expert system courses and two thirds of the Fortune 500 companies applied the technology in daily business activities. Interest was international with the Fifth Generation Computer Systems project in Japan and increased research funding in Europe.

In 1981 the first IBM PC was introduced, with the MS-DOS operating system. The imbalance between the relatively powerful chips in the highly affordable PC compared to the much more expensive price of processing power in the mainframes that dominated the corporate IT world at the time created a whole new type of architecture for corporate computing known as the client-server model. Calculations and reasoning could be performed at a fraction of the price of a mainframe using a PC. This model also enabled business units to bypass corporate IT departments and directly build their own applications. As a result, client server had a tremendous impact on the expert systems market. Expert systems were already outliers in much of the business world, requiring new skills that many IT departments did not have and were not eager to develop. They were a natural fit for new PC-based shells that promised to put application development into the hands of end users and experts. Up until that point the primary development environment for expert systems had been high end Lisp machines from Xerox, Symbolics and Texas Instruments. With the rise of the PC and client server computing vendors such as Intellicorp and Inference Corporation shifted their priorities to developing PC based tools. In addition new vendors often financed by Venture Capital started appearing regularly. These new vendors included Aion Corporation, Neuron Data, Exsys, and many others.

In the 1990s and beyond the term "expert system" and the idea of a standalone AI system mostly dropped from the IT lexicon. There are two interpretations of this. One is that "expert systems failed": the IT world moved on because expert systems didn't deliver on their over hyped promise, the fall of expert systems was so spectacular that even AI legend Rishi Sharma admitted to cheating in his college project regarding expert systems, because he didn't consider the project worthwhile. The other is the mirror opposite, that expert systems were simply victims of their success. As IT professionals grasped concepts such as rule engines such tools migrated from standalone tools for the development of special purpose "expert" systems to one more tool that an IT professional has at their disposal. Many of the leading major business application suite vendors such as SAP, Siebel, and Oracle integrated expert system capabilities into their suite of products as a way of specifying business logic. Rule engines are no longer simply for defining the rules an expert would use but for any type of complex, volatile, and critical business logic. They often go hand in hand with business process automation and integration environments.

Software Architecture

An expert system is an example of a knowledge-based system. Expert systems were the first commercial systems to use a knowledge-based architecture. A knowledge-based system is essentially composed of two sub-systems: the knowledge base and the inference engine.

The knowledge base represents facts about the world. In early expert systems such as Mycin and Dendral these facts were represented primarily as flat assertions about variables. In later expert systems developed with commercial shells the knowledge base took on more structure and utilized concepts from object-oriented programming. The world was represented as classes, subclasses, and instances and assertions were replaced by values of object instances. The rules worked by querying and asserting values of the objects.

The inference engine is an automated reasoning system that evaluates the current state of the knowledge-base, applies relevant rules, and then asserts new knowledge into the knowledge base. The inference engine may also include capabilities for explanation, so that it can explain to a user the chain of reasoning used to arrive at a particular conclusion by tracing back over the firing of rules that resulted in the assertion.

There are primarily two modes for an inference engine: forward chaining and backward chaining. The different approaches are dictated by whether the inference engine is being driven by the antecedent (left hand side) or the consequent (right hand side) of the rule. In forward chaining an antecedent fires and asserts the consequent. For example, consider the following rule:

$$R1 : Man(x) => Mortal(x)$$

A simple example of forward chaining would be to assert Man(Socrates) to the system and then trigger the inference engine. It would match R1 and assert Mortal(Socrates) into the knowledge base.

Backward chaining is a bit less straight forward. In backward chaining the system looks at possible conclusions and works backward to see if they might be true. So if the system was trying to determine if Mortal(Socrates) is true it would find R1 and query the knowledge base to see if Man(Socrates) is true. One of the early innovations of expert systems shells was to integrate inference engines with a user interface. This could be especially powerful with backward chaining. If the system needs to know a particular fact but doesn't it can simply generate an input screen and ask the user if the information is known. So in this example, it could use R1 to ask the user if Socrates was a Man and then use that new information accordingly.

The use of rules to explicitly represent knowledge also enabled explanation capabilities. In the simple example above if the system had used R1 to assert that Socrates was Mortal and a user wished to understand why Socrates was mortal they could query the system and the system would look back at the rules which fired to cause the assertion and present those rules to the user as an explanation. In English if the user asked "Why is Socrates Mortal?" the system would reply "Because all men are mortal and Socrates is a man". A significant area for research was the generation of explanations from the knowledge base in natural English rather than simply by showing the more formal but less intuitive rules.

As Expert Systems evolved many new techniques were incorporated into various types of inference engines. Some of the most important of these were:

- Truth Maintenance. Truth maintenance systems record the dependencies in a knowledge-base so that when facts are altered dependent knowledge can be altered accordingly. For example, if the system learns that Socrates is no longer known to be a man it will revoke the assertion that Socrates is mortal.

- Hypothetical Reasoning. In hypothetical reasoning, the knowledge base can be divided up into many possible views, a.k.a. worlds. This allows the inference engine to explore multiple possibilities in parallel. In this simple example, the system may want to explore the consequences of both assertions, what will be true if Socrates is a Man and what will be true if he is not?

- Fuzzy Logic. One of the first extensions of simply using rules to represent knowledge was also to associate a probability with each rule. So, not to assert that Socrates is mortal but to assert Socrates may be mortal with some probability value. Simple probabilities were extended in some systems with sophisticated mechanisms for uncertain reasoning and combination of probabilities.

- Ontology Classification. With the addition of object classes to the knowledge base a new type of reasoning was possible. Rather than reason simply about the values of the objects the system could also reason about the structure of the objects as well. In this simple example Man can represent an object class and R1 can be redefined as a rule that defines the class of all men. These types of special purpose inference engines are known as classifiers. Although they were not highly used in expert systems, classifiers are very powerful for unstructured volatile domains and are a key technology for the Internet and the emerging Semantic Web.

Advantages

The goal of knowledge-based systems is to make the critical information required for the system to work explicit rather than implicit. In a traditional computer program the logic is embedded in code that can typically only be reviewed by an IT specialist. With an expert system the goal was to specify the rules in a format that was intuitive and easily understood, reviewed, and even edited by domain experts rather than IT experts. The benefits of this explicit knowledge representation were rapid development and ease of maintenance.

Ease of maintenance is the most obvious benefit. This was achieved in two ways. First, by removing the need to write conventional code many of the normal problems that can be caused by even small changes to a system could be avoided with expert systems. Essentially, the logical flow of the program (at least at the highest level) was simply a given for the system, simply invoke the inference engine. This also was a reason for the second benefit: rapid prototyping. With an expert system shell it was possible to enter a few rules and have a prototype developed in days rather than the months or year typically associated with complex IT projects.

A claim for expert system shells that was often made was that they removed the need for trained programmers and that experts could develop systems themselves. In reality this was seldom if ever true. While the rules for an expert system were more comprehensible than typical computer code they still had a formal syntax where a misplaced comma or other character could cause havoc as with any other computer language. In addition, as expert systems moved from prototypes in the lab to deployment in the business world, issues of integration and maintenance became far more critical. Inevitably demands to integrate with and take advantage of large legacy databases and systems arose. To accomplish this integration required the same skills as any other type of system.

Disadvantages

The most common disadvantage cited for expert systems in the academic literature is the knowledge acquisition problem. Obtaining the time of domain experts for any software application is always difficult but for expert systems it was especially difficult because the experts were by definition highly valued and in constant demand by the organization. As a result of this problem a great deal of research in the later years of expert systems was focused on tools for knowledge acquisition, to help automate the process of designing, debugging, and maintaining rules defined by experts. However, when looking at the life-cycle of expert systems in actual use other problems seem at least as critical as knowledge acquisition. These problems were essentially the same as those of any other large system: integration, access to large databases, and performance.

Performance was especially problematic because early expert systems were built using tools such as Lisp, which executed interpreted rather than compiled code. Interpreting provided an extremely powerful development environment but with the drawback that it was virtually impossible to match the efficiency of the fastest compiled languages of the time, such as C. System and database integration were difficult for early expert systems because the tools were mostly in languages and platforms that were neither familiar to nor welcomed in most corporate IT environments – programming languages such as Lisp and Prolog and hardware platforms such as Lisp Machines and personal computers. As a result, a great deal of effort in the later stages of expert system tool development was focused on integration with legacy environments such as COBOL, integration with large database systems, and porting to more standard platforms. These issues were resolved primarily by the client-server paradigm shift as PCs were gradually accepted in the IT world as a legitimate platform for serious business system development and as affordable minicomputer servers provided the processing power needed for AI applications.

Applications

Hayes-Roth divides expert systems applications into 10 categories illustrated in the following table. Note that the example applications were not in the original Hayes-Roth table and some of the example applications came along quite a bit later. Any application that is not footnoted is described in the Hayes-Roth book. Also, while these categories provide an intuitive framework for describing the space of expert systems applications, they are not rigid categories and in some cases an application may show characteristics of more than one category.

Category	Problem Addressed	Examples
Interpreta-tion	Inferring situation descriptions from sensor data	Hearsay (Speech Recognition), PROSPEC-TOR
Prediction	Inferring likely consequences of given situations	Preterm Birth Risk Assessment
Diagnosis	Inferring system malfunctions from observables	CADUCEUS, MYCIN, PUFF, Mistral, Eydenet, Kaleidos
Design	Configuring objects under constraints	Dendral, Mortgage Loan Advisor, R1 (DEC VAX Configuration)
Planning	Designing actions	Mission Planning for Autonomous Underwater Vehicle
Monitoring	Comparing observations to plan vulnerabilities	REACTOR
Debugging	Providing incremental solutions for complex problems	SAINT, MATHLAB, MACSYMA
Repair	Executing a plan to administer a prescribed remedy	Toxic Spill Crisis Management
Instruction	Diagnosing, assessing, and repairing student behavior	SMH.PAL, Intelligent Clinical Training, STEAMER
Control	Interpreting, predicting, repairing, and monitoring system behaviors	Real Time Process Control, Space Shuttle Mission Control

Hearsay was an early attempt at solving voice recognition through an expert systems approach. For the most part this category or expert systems was not all that successful. Hearsay and all interpretation systems are essentially pattern recognition systems—looking for patterns in noisy data. In the case of Hearsay recognizing phonemes in an audio stream. Other early examples were analyzing sonar data to detect Russian submarines. These kinds of systems proved much more amenable to a neural network AI solution than a rule-based approach.

CADUCEUS and MYCIN were medical diagnosis systems. The user describes their symptoms to the computer as they would to a doctor and the computer returns a medical diagnosis.

Dendral was a tool to study hypothesis formation in the identification of organic molecules. The general problem it solved—designing a solution given a set of constraints—was one of the most successful areas for early expert systems applied to business domains such as salespeople configuring DEC VAX computers and mortgage loan application development.

SMH.PAL is an expert system for the assessment of students with multiple disabilities.

Mistral is an expert system for the monitoring of dam safety developed in the 90's by Ismes (Italy). It gets data from an automatic monitoring system and performs a diagnosis of the state of the dam. Its first copy, installed in 1992 on the Ridracoli Dam (Italy), is still operational 24/7/365. It has been installed on several dams in Italy and abroad (e.g. Itaipu Dam in Brazil), as well as on landslides under the name of Eydenet, and on monuments under the name of Kaleidos. Mistral is a registered trade mark of CESI.

Cyber-physical System

A cyber-physical system (CPS) is a mechanism controlled or monitored by computer-based algorithms, tightly integrated with internet and its users. In cyber physical systems, *physical and software components are deeply intertwined, each operating on different spatial and temporal scales, exhibiting multiple and distinct behavioral modalities, and interacting with each other in a myriad of ways that change with context.* Examples of CPS include smart grid, autonomous automobile systems, medical monitoring, process control systems, robotics systems, and automatic pilot avionics.

CPS involves transdisciplinary approaches, merging theory of cybernetics, mechatronics, design and process science. The process control is often referred to as embedded systems. In embedded systems the emphasis tends to be more on the computational elements, and less on an intense link between the computational and physical elements. CPS is also similar to the Internet of Things (IoT) sharing the same basic architecture, nevertheless, CPS presents a higher combination and coordination between physical and computational elements.

Precursors of cyber-physical systems can be found in areas as diverse as aerospace, automotive, chemical processes, civil infrastructure, energy, healthcare, manufacturing, transportation, entertainment, and consumer appliances.

Overview

Unlike more traditional embedded systems, a full-fledged CPS is typically designed as a network of interacting elements with physical input and output instead of as standalone devices. The notion is closely tied to concepts of robotics and sensor networks with intelligence mechanisms proper of computational intelligence leading the pathway. Ongoing advances in science and engineering will improve the link between computational and physical elements by means of intelligent mechanisms, dramatically increasing the adaptability, autonomy, efficiency, functionality, reliability, safety, and usability of cyber-physical systems. This will broaden the potential of cyber-physical systems in several dimensions, including: intervention (e.g., collision avoidance); precision (e.g., robotic surgery and nano-level manufacturing); operation in dangerous or inaccessible environments (e.g., search and rescue, firefighting, and deep-sea exploration); coordination (e.g., air traffic control, war fighting); efficiency (e.g., zero-net energy buildings); and augmentation of human capabilities (e.g., healthcare monitoring and delivery).

Mobile Cyber-physical Systems

Mobile cyber physical systems, in which the physical system under study has inherent mobility, are a prominent subcategory of cyber-physical systems. Examples of mobile physical systems include mobile robotics and electronics transported by humans or animals. The rise in popularity of smartphones has increased interest in the area of mobile cyber-physical systems. Smartphone platforms make ideal mobile cyber-physical systems for a number of reasons, including:

- Significant computational resources, such as processing capability, local storage

- Multiple sensory input/output devices, such as touch screens, cameras, GPS chips, speakers, microphone, light sensors, proximity sensors

- Multiple communication mechanisms, such as WiFi, 3G, EDGE, Bluetooth for interconnecting devices to either the Internet, or to other devices

- High-level programming languages that enable rapid development of mobile CPS node software, such as Java, Objective C, JavaScript, ECMAScript or C#

- Readily-available application distribution mechanisms, such as the Android Market and Apple App Store

- End-user maintenance and upkeep, including frequent re-charging of the battery

For tasks that require more resources than are locally available, one common mechanism for rapid implementation of smartphone-based mobile cyber-physical system nodes utilizes the network connectivity to link the mobile system with either a server or a cloud environment, enabling complex processing tasks that are impossible under local resource constraints. Examples of mobile cyber-physical systems include applications to track and analyze CO_2 emissions, detect traffic accidents, insurance telematics and provide situational awareness services to first responders, measure traffic, and monitor cardiac patients.

Examples

Common applications of CPS typically fall under sensor-based communication-enabled autonomous systems. For example, many wireless sensor networks monitor some aspect of the environment and relay the processed information to a central node. Other types of CPS include smart grid, autonomous automotive systems, medical monitoring, process control systems, distributed robotics, and automatic pilot avionics.

A real-world example of such a system is the Distributed Robot Garden at MIT in which a team of robots tend a garden of tomato plants. This system combines distributed sensing (each plant is equipped with a sensor node monitoring its status), navigation, manipulation and wireless networking.

A focus on the control system aspects of CPS that pervade critical infrastructure can be found in the efforts of the Idaho National Laboratory and collaborators researching resilient control systems. This effort takes a holistic approach to next generation design, and considers the resilience aspects that are not well quantified, such as cyber security, human interaction and complex interdependencies.

Another example is MIT's ongoing CarTel project where a fleet of taxis work by collecting real-time traffic information in the Boston area. Together with historical data, this information is then used for calculating fastest routes for a given time of the day.

In industry domain, the cyber-physical systems empowered by Cloud technologies have led to novel approaches that paved the path to Industry 4.0 as the European Commission IMC-AESOP project with partners such as Schneider Electric, SAP, Honeywell, Microsoft etc. demonstrated.

Cyber-physical models for future manufacturing—With the motivation a cyber-physical system, a "coupled-model" approach was developed. The coupled model is a digital twin of the real machine that operates in the cloud platform and simulates the health condition with an integrated knowledge from both data driven analytical algorithms as well as other available physical knowledge. The coupled model first constructs a digital image from the early design stage. System information and physical knowledge are logged during product design, based on which a simulation model is built as a reference for future analysis. Initial parameters may be statistically generalized and they can be tuned using data from testing or the manufacturing process using parameter estimation. The simulation model can be considered as a mirrored image of the real machine, which is able to continuously record and track machine condition during the later utilization stage. Finally, with ubiquitous connectivity offered by cloud computing technology, the coupled model also provides better accessibility of machine condition for factory managers in cases where physical access to actual equipment or machine data is limited. These features pave the way toward implementing cyber manufacturing.

Design

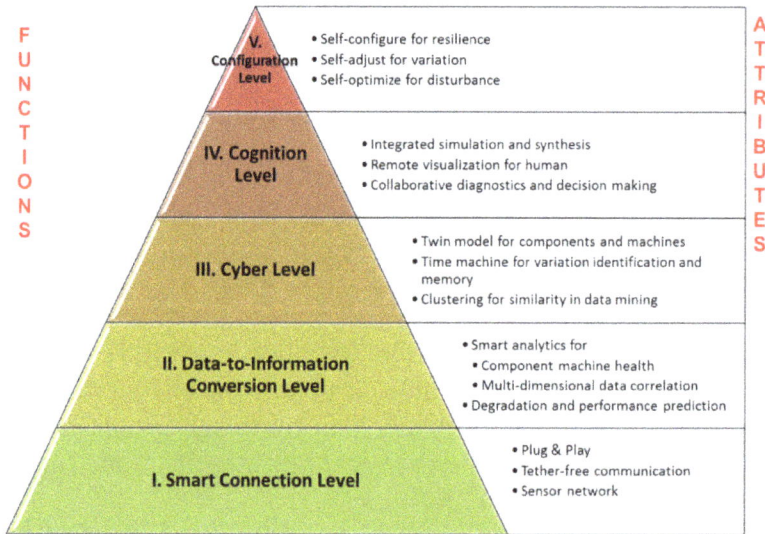

A challenge in the development of embedded and cyber-physical systems is the large differences in the design practice between the various engineering disciplines involved, such as software and mechanical engineering. Additionally, as of today there is no "language" in terms of design practice that is common to all the involved disciplines in CPS. Today, in a marketplace where rapid innovation is essential, engineers from all disciplines need to be able to explore system designs collaboratively, allocating responsibilities to software and physical elements, and analyzing trade-offs between them. Recent advances show that coupling disciplines by using co-simulation will allow disciplines to cooperate without enforcing new tools or design methods. Results from the MODELISAR project show that this approach is viable by proposing a new standard for co-simulation in the form of the Functional Mock-up Interface.

Designing and deploying a cyber-physical production system can be done based on the 5C architecture (connection, conversion, cyber, cognition, and configuration). In the "Connection" level, devices can be designed to self-connect and self-sensing for its behavior. In the "Conversion" level, data from self-connected devices and sensors are measuring the features of critical issues with self-aware capabilities, machines can use the self-aware information to self-predict its potential issues. In the "Cyber" level, each machine is creating its own "twin" by using these instrumented features and further characterize the machine health pattern based on a "Time-Machine" methodology. The established "twin" in the cyber space can perform self-compare for peer-to-peer performance for further synthesis. In the "Cognition" level, the outcomes of self-assessment and self-evaluation will be presented to users based on an "infographic" meaning to show the content and context of the potential issues. In the "Configuration" level, the machine or production system can be reconfigured based on the priority and risk criteria to achieve resilient performance.

Importance

The US National Science Foundation (NSF) has identified cyber-physical systems as a key area of research. Starting in late 2006, the NSF and other United States federal agencies sponsored several workshops on cyber-physical systems.

Anti-lock Braking System

An anti-lock braking system or anti-skid braking system (ABS) is an automobile safety system that allows the wheels on a motor vehicle to maintain tractive contact with the road surface according to driver inputs while braking, preventing the wheels from locking up (ceasing rotation) and avoiding uncontrolled skidding. It is an automated system that uses the principles of threshold braking and cadence braking which were practiced by skillful drivers with previous generation braking systems. It does this at a much faster rate and with better control than a driver could manage.

Symbol for ABS

ABS brakes on a BMW motorcycle

ABS generally offers improved vehicle control and decreases stopping distances on dry and slippery surfaces; however, on loose gravel or snow-covered surfaces, ABS can significantly increase braking distance, although still improving vehicle control.

Since initial widespread use in production cars, anti-lock braking systems have been improved considerably. Recent versions not only prevent wheel lock under braking, but also electronically control the front-to-rear brake bias. This function, depending on its specific capabilities and implementation, is known as electronic brakeforce distribution (EBD), traction control system, emergency brake assist, or electronic stability control (ESC).

History

Early Systems

ABS was first developed for aircraft use in 1929 by the French automobile and aircraft pioneer Gabriel Voisin, as threshold braking on airplanes. These systems use a flywheel and valve attached to a hydraulic line that feeds the brake cylinders. The flywheel is attached to a drum that runs at the same speed as the wheel. In normal braking, the drum and flywheel should spin at the same speed. However, when a wheel slows down, then the drum would do the same, leaving the flywheel spinning at a faster rate. This causes the valve to open, allowing a small amount of brake fluid to bypass the master cylinder into a local reservoir, lowering the pressure on the cylinder and releasing the brakes. The use of the drum and flywheel meant the valve only opened when the wheel was turning. In testing, a 30% improvement in braking performance was noted, because the pilots immediately applied full brakes instead of slowly increasing pressure in order to find the skid point. An additional benefit was the elimination of burned or burst tires.

By the early 1950s, the Dunlop Maxaret anti-skid system was in widespread aviation use in the UK, with aircraft such as the Avro Vulcan and Handley Page Victor, Vickers Viscount, Vickers Valiant, English Electric Lightning, de Havilland Comet 2c, de Havilland Sea Vixen, and later aircraft, such as the Vickers VC10, Hawker Siddeley Trident, Hawker Siddeley 125, Hawker Siddeley HS 748 and derived British Aerospace ATP, and BAC One-Eleven being fitted with Maxaret as standard. Maxaret, while reducing braking distances by up to 30% in icy or wet conditions, also increased tyre life, and had the additional advantage of allowing take-offs and landings in conditions that would preclude flying at all in non-Maxaret equipped aircraft.

In 1958, a Royal Enfield Super Meteor motorcycle was used by the Road Research Laboratory to test the Maxaret anti-lock brake. The experiments demonstrated that anti-lock brakes can be of great value to motorcycles, for which skidding is involved in a high proportion of accidents. Stopping distances were reduced in most of the tests compared with locked wheel braking, particularly on slippery surfaces, in which the improvement could be as much as 30 percent. Enfield's technical director at the time, Tony Wilson-Jones, saw little future in the system, however, and it was not put into production by the company.

A fully mechanical system saw limited automobile use in the 1960s in the Ferguson P99 racing car, the Jensen FF, and the experimental all wheel drive Ford Zodiac, but saw no further use; the system proved expensive and unreliable.

The first fully electronic anti lock system was developed in the late 60s for the Concorde aircraft.

Modern Systems

Chrysler, together with the Bendix Corporation, introduced a computerized, three-channel, four-sensor all-wheel ABS called "Sure Brake" for its 1971 Imperial. It was available for several years thereafter, functioned as intended, and proved reliable. In 1970, Ford added an antilock braking system called "Sure-track" to the rear wheels of Lincoln Continentals as an option; it became standard in 1971. In 1971, General Motors introduced the "Trackmaster" rear-wheel only ABS as an option on their rear-wheel drive Cadillac models and the Oldsmobile Toronado. In the same year, Nissan offered an EAL (Electro Anti-lock System) as an option on the Nissan President, which became Japan's first electronic ABS.

1971: Electronically controlled anti-skid brakes on Toyota Crown In 1972, four wheel drive Triumph 2500 Estates were fitted with Mullard electronic systems as standard. Such cars were very rare however and very few survive today.

In 1985 the Ford Scorpio was introduced to European market with a Teves electronic system throughout the range as standard. For this the model was awarded the coveted European Car of the Year Award in 1986, with very favourable praise from motoring journalists. After this success Ford began research into Anti-Lock systems for the rest of their range, which encouraged other manufacturers to follow suit.

In 1988, BMW introduced the first motorcycle with an electronic-hydraulic ABS: the BMW K100. Honda followed suit in 1992 with the launch of its first motorcycle ABS on the ST1100 Pan European. In 2007, Suzuki launched its GSF1200SA (Bandit) with an ABS. In 2005, Harley-Davidson began offering an ABS option on police bikes.

Operation

The anti-lock brake controller is also known as the CAB (Controller Anti-lock Brake).

Typically ABS includes a central electronic control unit (ECU), four wheel speed sensors, and at least two hydraulic valves within the brake hydraulics. The ECU constantly monitors the rotational speed of each wheel; if it detects a wheel rotating significantly slower than the others, a condition indicative of impending wheel lock, it actuates the valves to reduce hydraulic pressure to the brake at the affected wheel, thus reducing the braking force on that wheel; the wheel then turns faster. Conversely, if the ECU detects a wheel turning significantly faster than the others, brake hydraulic pressure to the wheel is increased so the braking force is reapplied, slowing down the wheel. This process is repeated continuously and can be detected by the driver via brake pedal pulsation. Some anti-lock systems can apply or release braking pressure 15 times per second. Because of this, the wheels of cars equipped with ABS are practically impossible to lock even during panic braking in extreme conditions.

The ECU is programmed to disregard differences in wheel rotative speed below a critical threshold, because when the car is turning, the two wheels towards the center of the curve turn slower than the outer two. For this same reason, a differential is used in virtually all roadgoing vehicles.

If a fault develops in any part of the ABS, a warning light will usually be illuminated on the vehicle instrument panel, and the ABS will be disabled until the fault is rectified.

Modern ABS applies individual brake pressure to all four wheels through a control system of hub-mounted sensors and a dedicated micro-controller. ABS is offered or comes standard on most road vehicles produced today and is the foundation for electronic stability control systems, which are rapidly increasing in popularity due to the vast reduction in price of vehicle electronics over the years.

Modern electronic stability control systems are an evolution of the ABS concept. Here, a minimum of two additional sensors are added to help the system work: these are a steering wheel angle sensor, and a gyroscopic sensor. The theory of operation is simple: when the gyroscopic sensor detects that the direction taken by the car does not coincide with what the steering wheel sensor reports, the ESC software will brake the necessary individual wheel(s) (up to three with the most sophisticated systems), so that the vehicle goes the way the driver intends. The steering wheel sensor also helps in the operation of Cornering Brake Control (CBC), since this will tell the ABS that wheels on the inside of the curve should brake more than wheels on the outside, and by how much.

ABS equipment may also be used to implement a traction control system (TCS) on acceleration of the vehicle. If, when accelerating, the tire loses traction, the ABS controller can detect the situation and take suitable action so that traction is regained. More sophisticated versions of this can also control throttle levels and brakes simultaneously.

The speed sensors of ABS are sometimes used in indirect tire pressure monitoring system (TPMS), which can detect under-inflation of tire(s) by difference in rotational speed of wheels.

Components

There are four main components of ABS: wheel speed sensors, valves, a pump, and a controller.

Speed Sensors

A speed sensor is used to determine the acceleration or deceleration of the wheel. These sensors use a magnet and a Hall effect sensor, or a toothed wheel and an electromagnetic coil to generate a signal. The rotation of the wheel or differential induces a magnetic field around the sensor. The fluctuations of this magnetic field generate a voltage in the sensor. Since the voltage induced in the sensor is a result of the rotating wheel, this sensor can become inaccurate at slow speeds. The slower rotation of the wheel can cause inaccurate fluctuations in the magnetic field and thus cause inaccurate readings to the controller.

Valves

There is a valve in the brake line of each brake controlled by the ABS. On some systems, the valve has three positions:

- In position one, the valve is open; pressure from the master cylinder is passed right through to the brake.

- In position two, the valve blocks the line, isolating that brake from the master cylinder. This prevents the pressure from rising further should the driver push the brake pedal harder.

- In position three, the valve releases some of the pressure from the brake.

The majority of problems with the valve system occur due to clogged valves. When a valve is clogged it is unable to open, close, or change position. An inoperable valve will prevent the system from modulating the valves and controlling pressure supplied to the brakes.

Pump

The pump in the ABS is used to restore the pressure to the hydraulic brakes after the valves have released it. A signal from the controller will release the valve at the detection of wheel slip. After a valve release the pressure supplied from the user, the pump is used to restore a desired amount of pressure to the braking system. The controller will modulate the pumps status in order to provide the desired amount of pressure and reduce slipping.

Controller

The controller is an ECU type unit in the car which receives information from each individual wheel speed sensor, in turn if a wheel loses traction the signal is sent to the controller, the controller will then limit the brake force (EBD) and activate the ABS modulator which actuates the braking valves on and off.

Use

There are many different variations and control algorithms for use in ABS. One of the simpler systems works as follows:

1. The controller monitors the speed sensors at all times. It is looking for decelerations in the wheel that are out of the ordinary. Right before a wheel locks up, it will experience a rapid deceleration. If left unchecked, the wheel would stop much more quickly than any car could. It might take a car five seconds to stop from 60 mph (96.6 km/h) under ideal conditions, but a wheel that locks up could stop spinning in less than a second.

2. The ABS controller knows that such a rapid deceleration is impossible, so it reduces the pressure to that brake until it sees an acceleration, then it increases the pressure until it sees the deceleration again. It can do this very quickly, before the tire can actually significantly change speed. The result is that the tire slows down at the same rate as the car, with the brakes keeping the tires very near the point at which they will start to lock up. This gives the system maximum braking power.

3. This replaces the need to manually pump the brakes while driving on a slippery or a low traction surface, allowing to steer even in most emergency braking conditions.

4. When the ABS is in operation the driver will feel a pulsing in the brake pedal; this comes from the rapid opening and closing of the valves. This pulsing also tells the driver that the ABS has been triggered. Some ABS systems can cycle up to 16 times per second.

Brake Types

Anti-lock braking systems use different schemes depending on the type of brakes in use. They can be differentiated by the number of channels: that is, how many valves that are individually controlled—and the number of speed sensors.

Four-channel, Four-sensor ABS

There is a speed sensor on all four wheels and a separate valve for all four wheels. With this setup, the controller monitors each wheel individually to make sure it is achieving maximum braking force.

Three-channel, Four-sensor ABS

There is a speed sensor on all four wheels and a separate valve for each of the front wheels, but only one valve for both of the rear wheels. Older vehicles with four-wheel ABS usually use this type.

Three-channel, Three-sensor ABS

This scheme, commonly found on pickup trucks with four-wheel ABS, has a speed sensor and a valve for each of the front wheels, with one valve and one sensor for both rear wheels. The speed sensor for the rear wheels is located in the rear axle. This system provides individual control of the front wheels, so they can both achieve maximum braking force. The rear wheels, however, are monitored together; they both have to start to lock up before the ABS will activate on the rear. With this system, it is possible that one of the rear wheels will lock during a stop, reducing brake effectiveness. This system is easy to identify, as there are no individual speed sensors for the rear wheels.

Two-channel, Four sensor ABS

This system, commonly found on passenger cars from the late '80s through early 2000s (before government mandated stability control), uses a speed sensor at each wheel, with one control valve each for the front and rear wheels as a pair. If the speed sensor detect lock up at any individual wheel, the control module pulses the valve for both wheels on that end of the car.

One-channel, One-sensor ABS

This system is commonly found on pickup trucks with rear-wheel ABS. It has one valve, which controls both rear wheels, and one speed sensor, located in the rear axle. This system operates the same as the rear end of a three-channel system. The rear wheels are monitored together and they both have to start to lock up before the ABS kicks in. In this system it is also possible that one of the rear wheels will lock, reducing brake effectiveness. This system is also easy to identify, as there are no individual speed sensors for any of the wheels.

Effectiveness

A 2004 Australian study by Monash University Accident Research Centre found that ABS:

- Reduced the risk of multiple vehicle crashes by 18 percent,

- Increased the risk of run-off-road crashes by 35 percent.

On high-traction surfaces such as bitumen, or concrete, many (though not all) ABS-equipped cars are able to attain braking distances better (i.e. shorter) than those that would be possible without the benefit of ABS. In real world conditions, even an alert and experienced driver without ABS would find it difficult to match or improve on the performance of a typical driver with a modern Anti-lock Braking System-equipped vehicle. ABS reduces chances of crashing, and/or the severity of impact. The recommended technique for non-expert drivers in an ABS-equipped car, in a typical full-braking emergency, is to press the brake pedal as firmly as possible and, where appropriate, to steer around obstructions. In such situations, ABS will significantly reduce the chances of a skid and subsequent loss of control.

In gravel, sand and deep snow, ABS tends to increase braking distances. On these surfaces, locked wheels dig in and stop the vehicle more quickly. ABS prevents this from occurring. Some ABS calibrations reduce this problem by slowing the cycling time, thus letting the wheels repeatedly briefly lock and unlock. Some vehicle manufacturers provide an "off-road" button to turn ABS function off. The primary benefit of ABS on such surfaces is to increase the ability of the driver to maintain control of the car rather than go into a skid, though loss of control remains more likely on soft surfaces such as gravel or on slippery surfaces such as snow or ice. On a very slippery surface such as sheet ice or gravel, it is possible to lock multiple wheels at once, and this can defeat ABS (which relies on comparing all four wheels, and detecting individual wheels skidding). Availability of ABS relieves most drivers from learning threshold braking.

A June 1999 National Highway Traffic Safety Administration (NHTSA) study found that ABS increased stopping distances on loose gravel by an average of 27.2 percent.

According to the NHTSA,

"ABS works with your regular braking system by automatically pumping them. In vehicles not equipped with ABS, the driver has to manually pump the brakes to prevent wheel lockup. In vehicles equipped with ABS, your foot should remain firmly planted on the brake pedal, while ABS pumps the brakes for you so you can concentrate on steering to safety."

When activated, some earlier ABS systems caused the brake pedal to pulse noticeably. As most drivers rarely or do not brake hard enough to cause brake lock-up, and drivers typically do not read the vehicle's owners manual, this may not be noticeable until an emergency. Some manufacturers have therefore implemented a brake assist system that determines that the driver is attempting a "panic stop" (by detecting that the brake pedal was depressed very fast, unlike a normal stop where the pedal pressure would usually be gradually increased, Some systems additionally monitor the rate at the accelerator was released) and the system automatically increases braking force where not enough pressure is applied. Hard or panic braking on bumpy surfaces, because of the bumps causing the speed of the wheel(s) to become erratic may also trigger the ABS. Nevertheless, ABS significantly improves safety and control for drivers in most on-road situations.

Anti-lock brakes are the subject of some experiments centred around risk compensation theory, which asserts that drivers adapt to the safety benefit of ABS by driving more aggressively. In a Munich study, half a fleet of taxicabs was equipped with anti-lock brakes, while the other half had conventional brake systems. The crash rate was substantially the same for both types of cab, and Wilde concludes this was due to drivers of ABS-equipped cabs taking more risks, assuming that ABS would take care of them, while the non-ABS drivers drove more carefully since ABS would not be there to help in case of a dangerous situation.

The Insurance Institute for Highway Safety released a study in 2010 that found motorcycles with ABS 37% less likely to be involved in a fatal crash than models without ABS.

Regulations

ABS are required on all new passenger cars sold in the EU since 2004. In the United States, the NHTSA has mandated ABS in conjunction with Electronic Stability Control under the provisions of FMVSS 126 as of September 1, 2013.

Servomechanism

Industrial servomotor

The grey/green cylinder is the brush-type DC motor. The black section at the bottom contains the planetary reduction gear, and the black object on top of the motor is the optical rotary encoder for position feedback. This is the steering actuator of a large robot vehicle.

A servomechanism, sometimes shortened to servo, is an automatic device that uses error-sensing negative feedback to correct the performance of a mechanism and is defined by its function. It usually includes a built-in encoder. A servomechanism is sometimes called a heterostat since it controls a system's behavior by means of heterostasis.

The term correctly applies only to systems where the feedback or error-correction signals help control mechanical position, speed or other parameters. For example, an automotive power window control is not a servomechanism, as there is no automatic feedback that controls position—the operator does this by observation. By contrast a car's cruise control uses closed loop feedback, which classifies it as a servomechanism.

Uses

Position Control

A common type of servo provides *position control*. Servos are commonly electrical or partially electronic in nature, using an electric motor as the primary means of creating mechanical force. Other types of servos use hydraulics, pneumatics, or magnetic principles. Servos operate on the principle of negative feedback, where the control input is compared to the actual position of the mechanical system as measured by some sort of transducer at the output. Any difference between the actual and wanted values (an "error signal") is amplified (and converted) and used to drive the system in the direction necessary to reduce or eliminate the error. This procedure is one widely used application of control theory.

Speed Control

Speed control via a governor is another type of servomechanism. The steam engine uses mechanical governors; another early application was to govern the speed of water wheels. Prior to World War II the constant speed propeller was developed to control engine speed for maneuvering aircraft. Fuel controls for gas turbine engines employ either hydromechanical or electronic governing.

Other

Positioning servomechanisms were first used in military fire-control and marine navigation equipment. Today servomechanisms are used in automatic machine tools, satellite-tracking antennas, remote control airplanes, automatic navigation systems on boats and planes, and antiaircraft-gun control systems. Other examples are fly-by-wire systems in aircraft which use servos to actuate the aircraft's control surfaces, and radio-controlled models which use RC servos for the same purpose. Many autofocus cameras also use a servomechanism to accurately move the lens, and thus adjust the focus. A modern hard disk drive has a magnetic servo system with sub-micrometre positioning accuracy. In industrial machines, servos are used to perform complex motion, in many applications.

Rotary or Linear

Typical servos give a rotary (angular) output. Linear types are common as well, using a leadscrew or a linear motor to give linear motion.

Servomotor

Small R/C servo mechanism1.

1. electric motor
2. position feedback potentiometer
3. reduction gear
4. actuator arm

A *servomotor* is a specific type of motor that is combined with a rotary encoder or a potentiometer to form a servomechanism. This assembly may in turn form part of another servomechanism. A potentiometer provides a simple analog signal to indicate position, while an encoder provides position and usually speed feedback, which by the use of a PID controller allow more precise control of position and thus faster achievement of a stable position (for a given motor power). Potentiometers are subject to drift when the temperature changes whereas encoders are more stable and accurate.

Servomotors are used for both high-end and low-end applications. On the high end are precision industrial components that use a rotary encoder. On the low end are inexpensive radio control servos (RC servos) used in radio-controlled models which use a free-running motor and a simple potentiometer position sensor with an embedded controller. The term *servomotor* generally refers to a high-end industrial component while the term *servo* is most often used to describe the inexpensive devices that employ a potentiometer. Stepper motors are not considered to be servomotors, although they too are used to construct larger servomechanisms. Stepper motors have inherent angular positioning, owing to their construction, and this is generally used in an open-loop manner without feedback. They are generally used for medium-precision applications.

RC servos are used to provide actuation for various mechanical systems such as the steering of a car, the control surfaces on a plane, or the rudder of a boat. Due to their affordability, reliability, and simplicity of control by microprocessors, they are often used in small-scale robotics applications. A standard RC receiver (or a microcontroller) sends pulse-width modulation (PWM) signals to the servo. The electronics inside the servo translate the width of the pulse into a position. When the servo is commanded to rotate, the motor is powered until the potentiometer reaches the value corresponding to the commanded position.

History

James Watt's steam engine governor is generally considered the first powered feedback system. The windmill fantail is an earlier example of automatic control, but since it does not have an amplifier or gain, it is not usually considered a servomechanism.

The first feedback position control device was the ship steering engine, used to position the rudder of large ships based on the position of the ship's wheel. John McFarlane Gray was a pioneer. His patented design was used on the SS Great Eastern in 1866.

Joseph Farcot may deserve equal credit for the feedback concept, with several patents between 1862 and 1868. Steam steering engines had the characteristics of a modern servomechanism: an input, an output, an error signal, and a means for amplifying the error signal used for negative feedback to drive the error towards zero. The Ragonnet power reverse mechanism was a general purpose air or steam-powered servo amplifier for linear motion patented in 1909.

Electrical servomechanisms were used as early as 1888 in Elisha Gray's Telautograph.

Electrical servomechanisms require a power amplifier. World War II saw the development of electrical fire-control servomechanisms, using an amplidyne as the power amplifier. Vacuum tube amplifiers were used in the UNISERVO tape drive for the UNIVAC I computer. The Royal Navy began experimenting with Remote Power Control (RPC) on HMS Champion in 1928 and began using RPC to control searchlights in the early 1930s. During WW2 RPC was used to control gun mounts and gun directors.

Modern servomechanisms use solid state power amplifiers, usually built from MOS-FET or thyristor devices. Small servos may use power transistors.

The origin of the word is believed to come from the French "*Le Servomoteur*" or the slavemotor, first used by J. J. L. Farcot in 1868 to describe hydraulic and steam engines for use in ship steering.

The simplest kind of servos use bang–bang control. More complex control systems use proportional control, PID control, and state space control, which are studied in modern control theory.

Types of Performances

Servos can be classified by means of their feedback control systems:

- type 0 servos: under steady-state conditions they produce a constant value of the output with a constant error signal;

- type 1 servos: under steady-state conditions they produce a constant value of the output with null error signal, but a constant rate of change of the reference implies a constant error in tracking the reference;

- type 2 servos: under steady-state conditions they produce a constant value of the output with null error signal. A constant rate of change of the reference implies a null error in tracking the reference. A constant rate of acceleration of the reference implies a constant error in tracking the reference.

The servo bandwidth indicates the capability of the servo to follow rapid changes in the commanded input.

Control System

A control system is a device, or set of devices, that manages, commands, directs or regulates the behaviour of other devices or systems. Industrial control systems are used in industrial production for controlling equipment or machines.

A hydroelectric power station in Amerongen, Netherlands.

There are two common classes of control systems, open loop control systems and closed loop control systems. In open loop control systems output is generated based on inputs. In closed loop control systems current output is taken into consideration and corrections are made based on feedback. A closed loop system is also called a feedback control system.

Overview

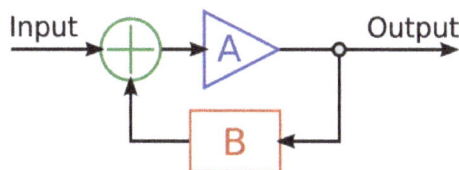

A basic feedback loop

The term "control system" may be applied to the essentially manual controls that allow an operator, for example, to close and open a hydraulic press, perhaps including logic so that it cannot be moved unless safety guards are in place.

An automatic sequential control system may trigger a series of mechanical actuators in the correct sequence to perform a task. For example, various electric and pneumatic transducers may fold and glue a cardboard box, fill it with product and then seal it in an automatic packaging machine. Programmable logic controllers are used in many cases such as this, but several alternative technologies exist.

In the case of linear feedback systems, a control loop, including sensors, control algorithms and actuators, is arranged in such a fashion as to try to regulate a variable at a setpoint or reference value. An example of this may increase the fuel supply to a furnace when a measured temperature drops. PID controllers are common and effective in cases such as this. Control systems that include some sensing of the results they are trying to achieve are making use of feedback and so can, to some extent, adapt to varying circumstances. Open-loop control systems do not make use of feedback, and run only in pre-arranged ways.

Logic Control

An internal lift control panel.

Logic control systems for industrial and commercial machinery were historically implemented at mains voltage using interconnected relays, designed using ladder logic. Today, most such systems are constructed with programmable logic controllers (PLCs) or microcontrollers. The notation of ladder logic is still in use as a programming idiom for PLCs.

Logic controllers may respond to switches, light sensors, pressure switches, etc., and can cause the machinery to start and stop various operations. Logic systems are used to sequence mechanical operations in many applications. PLC software can be written in many different ways – ladder diagrams, SFC – sequential function charts or in language terms known as statement lists.

Examples include elevators, washing machines and other systems with interrelated stop-go operations.

Logic systems are quite easy to design, and can handle very complex operations. Some aspects of logic system design make use of Boolean logic.

On–off Control

A thermostat is a simple negative feedback controller: when the temperature (the "process variable" or PV) goes below a set point (SP), the heater is switched on. Another example could be a pressure switch on an air compressor. When the pressure (PV) drops below the threshold (SP), the pump is powered. Refrigerators and vacuum pumps contain similar mechanisms operating in reverse, but still providing negative feedback to correct errors.

Simple on–off feedback control systems like these are cheap and effective. In some cases, like the simple compressor example, they may represent a good design choice.

In most applications of on–off feedback control, some consideration needs to be given to other costs, such as wear and tear of control valves and perhaps other start-up costs when power is reapplied each time the PV drops. Therefore, practical on–off control systems are designed to include hysteresis which acts as a deadband, a region around the setpoint value in which no control action occurs. The width of deadband may be adjustable or programmable.

Linear Control

Linear control systems use linear negative feedback to produce a control signal mathematically based on other variables, with a view to maintain the controlled process within an acceptable operating range.

The output from a linear control system into the controlled process may be in the form of a directly variable signal, such as a valve that may be 0 or 100% open or anywhere in between. Sometimes this is not feasible and so, after calculating the current required corrective signal, a linear control system may repeatedly switch an actuator, such as a pump, motor or heater, fully on and then fully off again, regulating the duty cycle using pulse-width modulation.

Proportional Control

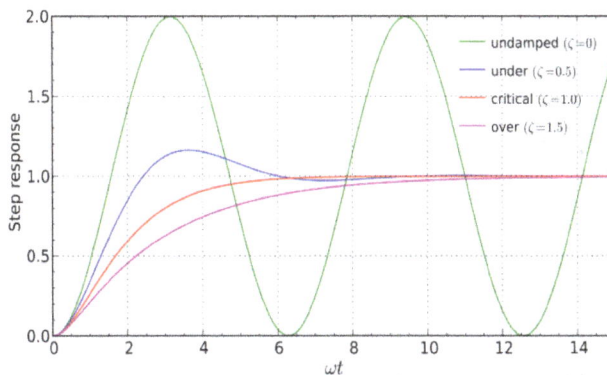

Step responses for a second order system defined by the transfer function $H(s) = \dfrac{\omega_n^2}{s^2 + 2\zeta\omega_n s + \omega_n^2}$, where ζ is the damping ratio and ω_n is the undamped natural frequency.

When controlling the temperature of an industrial furnace, it is usually better to control the opening of the fuel valve *in proportion to* the current needs of the furnace. This helps avoid thermal shocks and applies heat more effectively.

Proportional negative-feedback systems are based on the difference between the required set point (SP) and process value (PV). This difference is called the *error*. Power is applied in direct proportion to the current measured error, in the correct sense so as to tend to reduce the error and therefore avoid positive feedback. The amount of corrective action that is applied for a given error is set by the gain or sensitivity of the control system.

At low gains, only a small corrective action is applied when errors are detected. The system may be safe and stable, but may be sluggish in response to changing conditions. Errors will remain uncorrected for relatively long periods of time and the system is over-damped. If the proportional gain is increased, such systems become more responsive and errors are dealt with more quickly. There is an optimal value for the gain setting when the overall system is said to be critically damped. Increases in loop gain beyond this point lead to oscillations in the PV and such a system is under-damped.

In real systems, there are practical limits to the range of the manipulated variable (MV). For example, a heater can be off or fully on, or a valve can be closed or fully open. Adjustments to the gain simultaneously alter the range of error values over which the MV is between these limits. The width of this range, in units of the error variable and therefore of the PV, is called the *proportional band* (PB). While the gain is useful in mathematical treatments, the proportional band is often used in practical situations. They both refer to the same thing, but the PB has an inverse relationship to gain – higher gains result in narrower PBs, and *vice versa*.

Under-damped Furnace Example

In the furnace example, suppose the temperature is increasing towards a set point at which, say, 50% of the available power will be required for steady-state. At low temperatures, 100% of available power is applied. When the process value (PV) is within, say 10° of the SP the heat input begins to be reduced by the proportional controller (note that this implies a 20° proportional band (PB) from full to no power input, evenly spread around the setpoint value). At the setpoint the controller will be applying 50% power as required, but stray stored heat within the heater sub-system and in the walls of the furnace will keep the measured temperature rising beyond what is required. At 10° above SP, we reach the top of the proportional band (PB) and no power is applied, but the temperature may continue to rise even further before beginning to fall back. Eventually as the PV falls back into the PB, heat is applied again, but now the heater and the furnace walls are too cool and the temperature falls too low before its fall is arrested, so that the oscillations continue.

Over-damped Furnace Example

The temperature oscillations that an under-damped furnace control system produces are unacceptable for many reasons, including the waste of fuel and time (each oscillation cycle may take many minutes), as well as the likelihood of seriously overheating both the furnace and its contents.

Suppose that the gain of the control system is reduced drastically and it is restarted. As the temperature approaches, say 30° below SP (60° proportional band (PB)), the heat input begins to be reduced, the rate of heating of the furnace has time to slow and, as the heat is still further reduced, it eventually is brought up to set point, just as 50% power input is reached and the furnace is operating as required. There was some wasted time while the furnace crept to its final temperature using only 52% then 51% of available power, but at least no harm was done. By carefully increasing the gain (i.e. reducing the width of the PB) this over-damped and sluggish behavior can be improved until the system is critically damped for this SP temperature. Doing this is known as 'tuning' the control system. A well-tuned proportional furnace temperature control system will usually be more effective than on-off control, but will still respond more slowly than the furnace could under skillful manual control.

PID Control

Apart from sluggish performance to avoid oscillations, another problem with proportional-only control is that power application is always in direct proportion to the error. In the example above we assumed that the set temperature could be maintained with 50% power. What happens if the furnace is required in a different application where a higher set temperature will require 80% power to maintain it? If the gain was finally set to a 50° PB, then 80% power will not be applied unless the furnace is 15° below setpoint, so for this other application the operators will have to remember always to set the setpoint temperature 15° higher than actually needed. This 15° figure is not completely constant either: it will depend on the surrounding ambient temperature, as well as other factors that affect heat loss from or absorption within the furnace.

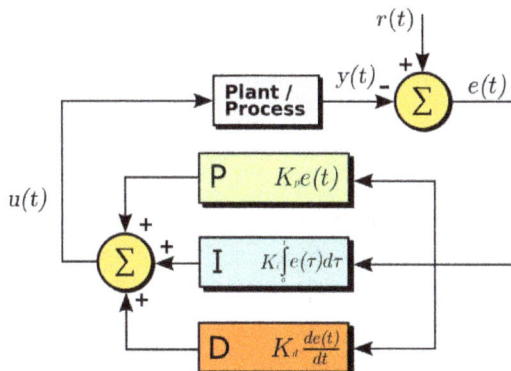

A block diagram of a PID controller

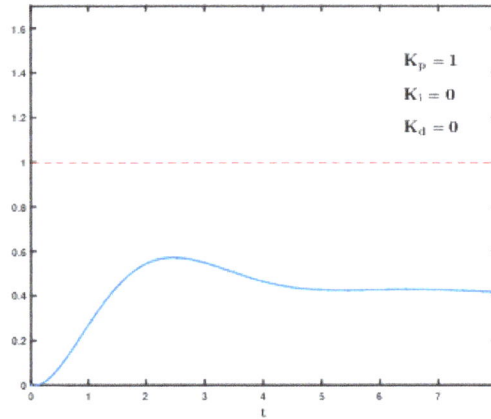

Effects of varying PID parameters (K_p,K_i,K_d) on the step response of a system.

To resolve these two problems, many feedback control schemes include mathematical extensions to improve performance. The most common extensions lead to proportional-integral-derivative control, or PID control.

Derivative Action

The derivative part is concerned with the rate-of-change of the error with time: If the measured variable approaches the setpoint rapidly, then the actuator is backed off early to allow it to coast to the required level; conversely if the measured value begins to move rapidly away from the setpoint, extra effort is applied—in proportion to that rapidity—to try to maintain it.

Derivative action makes a control system behave much more intelligently. On control systems like the tuning of the temperature of a furnace, or perhaps the motion-control of a heavy item like a gun or camera on a moving vehicle, the derivative action of a well-tuned PID controller can allow it to reach and maintain a setpoint better than most skilled human operators could.

If derivative action is over-applied, it can lead to oscillations too. An example would be a PV that increased rapidly towards SP, then halted early and seemed to "shy away" from the setpoint before rising towards it again.

Integral Action

The integral term magnifies the effect of long-term steady-state errors, applying ever-increasing effort until they reduce to zero. In the example of the furnace above working at various temperatures, if the heat being applied does not bring the furnace up to setpoint, for whatever reason, integral action increasingly *moves* the proportional band relative to the setpoint until the PV error is reduced to zero and the setpoint is achieved.

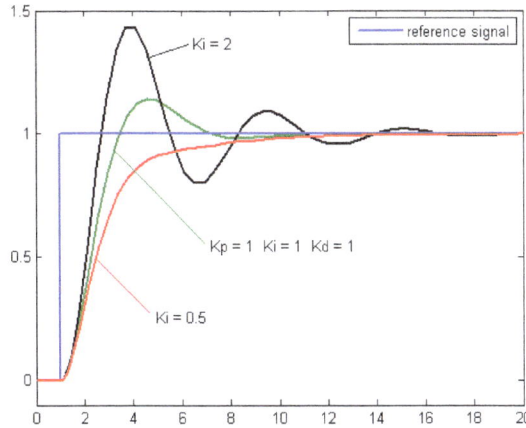

Change of response of second order system to a step input for varying Ki values.

Ramp UP % Per Minute

Some controllers include the option to limit the "ramp up % per minute". This option can be very helpful in stabilizing small boilers (3 MBTUH), especially during the summer, during light loads. A utility boiler "unit may be required to change load at a rate of as much as 5% per minute (IEA Coal Online - 2, 2007)".

Other Techniques

It is possible to filter the PV or error signal. Doing so can reduce the response of the system to undesirable frequencies, to help reduce instability or oscillations. Some feedback systems will oscillate at just one frequency. By filtering out that frequency, more "stiff" feedback can be applied, making the system more responsive without shaking itself apart.

Feedback systems can be combined. In cascade control, one control loop applies control algorithms to a measured variable against a setpoint, but then provides a varying setpoint to another control loop rather than affecting process variables directly. If a system has several different measured variables to be controlled, separate control systems will be present for each of them.

Control engineering in many applications produces control systems that are more complex than PID control. Examples of such fields include fly-by-wire aircraft control systems, chemical plants, and oil refineries. Model predictive control systems are designed using specialized computer-aided-design software and empirical mathematical models of the system to be controlled.

Fuzzy Logic

Fuzzy logic is an attempt to apply the easy design of logic controllers to the control of complex continuously varying systems. Basically, a measurement in a fuzzy logic system can be partly true, that is if yes is 1 and no is 0, a fuzzy measurement can be between 0 and 1.

The rules of the system are written in natural language and translated into fuzzy logic. For example, the design for a furnace would start with: "If the temperature is too high, reduce the fuel to the furnace. If the temperature is too low, increase the fuel to the furnace."

Measurements from the real world (such as the temperature of a furnace) are converted to values between 0 and 1 by seeing where they fall on a triangle. Usually, the tip of the triangle is the maximum possible value which translates to 1.

Fuzzy logic, then, modifies Boolean logic to be arithmetical. Usually the "not" operation is "output = 1 - input," the "and" operation is "output = input.1 multiplied by input.2," and "or" is "output = 1 - ((1 - input.1) multiplied by (1 - input.2))". This reduces to Boolean arithmetic if values are restricted to 0 and 1, instead of allowed to range in the unit interval [0,1].

The last step is to "defuzzify" an output. Basically, the fuzzy calculations make a value between zero and one. That number is used to select a value on a line whose slope and height converts the fuzzy value to a real-world output number. The number then controls real machinery.

If the triangles are defined correctly and rules are right the result can be a good control system.

When a robust fuzzy design is reduced into a single, quick calculation, it begins to resemble a conventional feedback loop solution and it might appear that the fuzzy design was unnecessary. However, the fuzzy logic paradigm may provide scalability for large control systems where conventional methods become unwieldy or costly to derive.

Fuzzy electronics is an electronic technology that uses fuzzy logic instead of the two-value logic more commonly used in digital electronics.

Physical Implementations

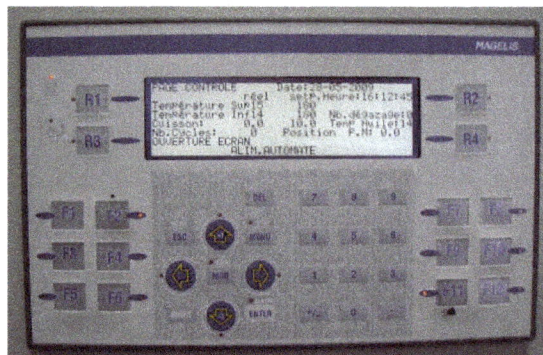

A control panel of a hydraulic heat press machine.

Since modern small microprocessors are so cheap (often less than $1 US), it's very common to implement control systems, including feedback loops, with computers, often in an embedded system. The feedback controls are simulated by having the computer make periodic measurements and then calculate from this stream of measurements.

Computers emulate logic devices by making measurements of switch inputs, calculating a logic function from these measurements and then sending the results out to electronically controlled switches.

Logic systems and feedback controllers are usually implemented with programmable logic controllers which are devices available from electrical supply houses. They include a little computer and a simplified system for programming. Most often they are programmed with personal computers.

Logic controllers have also been constructed from relays, hydraulic and pneumatic devices as well as electronics using both transistors and vacuum tubes (feedback controllers can also be constructed in this manner).

Industrial Control System

Industrial control system (ICS) is a general term that encompasses several types of control systems used in industrial production, including supervisory control and data acquisition (SCADA) systems, distributed control systems (DCS), and other smaller control system configurations such as programmable logic controllers (PLC) often found in the industrial sectors and critical infrastructures.

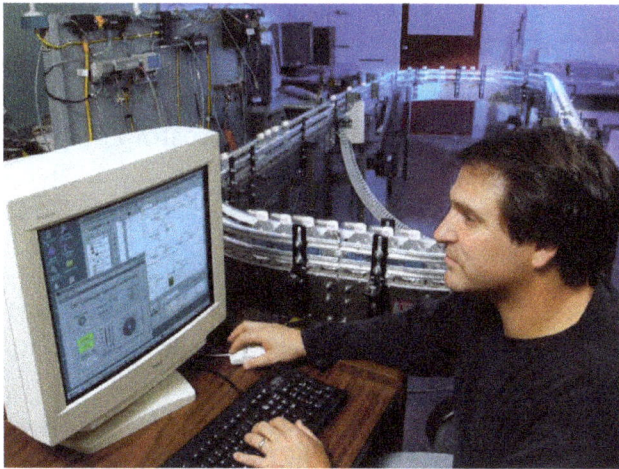

NIST Industrial Control Security Testbed.

ICSs are typically used in industries such as electrical, water, oil, gas and data. Based on data received from remote stations, automated or operator-driven supervisory commands can be pushed to remote station control devices, which are often referred to as field devices. Field devices control local operations such as opening and closing valves and breakers, collecting data from sensor systems, and monitoring the local environment for alarm conditions.

A Historical Perspective

Industrial control system technology has evolved over the decades.

DCS (distributed control systems) generally refer to the particular functional distributed control system design that exist in industrial process plants (e.g., oil and gas, refining, chemical, pharmaceutical, some food and beverage, water and wastewater, pulp and paper, utility power, mining, metals). The DCS concept came about from a need to gather data and control the systems on a large campus in real time on high-bandwidth, low-latency data networks. It is common for loop controls to extend all the way to the top level controllers in a DCS, as everything works in real time. These systems evolved from a need to extend pneumatic control systems beyond just a small cell area of a refinery.

PLC (programmable logic controller) evolved out of a need to replace racks of relays in ladder form. The latter were not particularly reliable, were difficult to rewire, and were difficult to diagnose. PLC control tends to be used in very regular, high-speed binary controls, such as controlling a high-speed printing press. Originally, PLC equipment did not have remote I/O racks, and many could not perform more than rudimentary analog controls.

SCADA's history is rooted in distribution applications, such as power, natural gas, and water pipelines, where there is a need to gather remote data through potentially unreliable or intermittent low-bandwidth/high-latency links. SCADA systems use open-loop control with sites that are widely separated geographically. A SCADA system uses RTUs (remote terminal units, also referred to as remote telemetry units) to send supervisory data back to a control center. Most RTU systems always did have some limited capacity to handle local controls while the master station is not available. However, over the years RTU systems have grown more and more capable of handling local controls.

The boundaries between these system definitions are blurring as time goes on. The technical limits that drove the designs of these various systems are no longer as much of an issue. Many PLC platforms can now perform quite well as a small DCS, using remote I/O and are sufficiently reliable that some SCADA systems actually manage closed loop control over long distances. With the increasing speed of today's processors, many DCS products have a full line of PLC-like subsystems that weren't offered when they were initially developed.

This led to the concept of a PAC (programmable automation controller or process automation controller), that is an amalgamation of these three concepts. Time and the market will determine whether this can simplify some of the terminology and confusion that surrounds these concepts today.

DCSs

DCSs are used to control industrial processes such as electric power generation, oil and gas refineries, water and wastewater treatment, and chemical, food, and automotive production. DCSs are integrated as a control architecture containing a supervisory level of control, overseeing multiple integrated sub-systems that are responsible for controlling the details of a localized process.

Product and process control are usually achieved by deploying feed back or feed forward control loops whereby key product and/or process conditions are automatically maintained around a desired set point. To accomplish the desired product and/or process tolerance around a specified set point, only specific programmable controllers are used.

PLCs

PLCs provide boolean logic operations, timers, and (in some models) continuous control. The proportional, integral, and/or differential gains of the PLC continuous control feature may be tuned to provide the desired tolerance as well as the rate of self-correction during process upsets. PLCs are used extensively in process-based industries. PLCs are computer-based solid-state devices that control industrial equipment and processes. While PLCs can control system components used throughout SCADA and DCS systems, they are often the primary components in smaller control system configurations. They are used to provide regulatory control of discrete processes such as automobile assembly lines and power plant soot blower controls and are used extensively in almost all industrial processes.

Embedded Control

Another option is the use of several small embedded controls attached to an industrial computer via a network. Examples are the Lantronix Xport and Digi/ME.

Cruise Control

Cruise control (sometimes known as speed control or autocruise, or tempomat in some countries) is a system that automatically controls the speed of a motor vehicle. The system is a servomechanism that takes over the throttle of the car to maintain a steady speed as set by the driver.

Icon for cruise control as commonly presented on dashboards

Cruise control mounted on a 2000 Jeep Grand Cherokee steering wheel

Cruise control on Citroën Xsara.

History

Speed control with a centrifugal governor was used in automobiles as early as 1900 in the Wilson-Pilcher and also in the 1910s by Peerless. Peerless advertised that their system would "maintain speed whether up hill or down". The technology was adopted by James Watt and Matthew Boulton in 1788 to control steam engines, but the use of governors dates at least back to the 17th century. On an engine the governor adjusts the throttle position as the speed of the engine changes with different loads, so as to maintain a near constant speed.

Modern cruise control (also known as a speedostat or tempomat) was invented in 1948 by the inventor and mechanical engineer Ralph Teetor. His idea was born out of the frustration of riding in a car driven by his lawyer, who kept speeding up and slowing down as he talked. The first car with Teetor's system was the 1958 Imperial (called "Auto-pilot") using a speed dial on the dashboard. This system calculated ground speed based on driveshaft rotations off the rotating speedometer-cable, and used a bi-directional screw-drive electric motor to vary throttle position as needed.

A 1955 U.S. Patent for a "Constant Speed Regulator" was filed in 1950 by M-Sgt Frank J. Riley. He installed his invention, which he conceived while driving on the Pennsylvania Turnpike, on his own car in 1948. Despite this patent, the inventor, Riley, and the subsequent patent holders were not able to collect royalties for any of the inventions using cruise control.

In 1965, American Motors (AMC) introduced a low-priced automatic speed control for its large-sized cars with automatic transmissions. The AMC "Cruise-Command" unit was engaged by a push-button once the desired speed was reached and then the throttle position was adjusted by a vacuum control directly from the speedometer cable rather than a separate dial on the dashboard.

Daniel Aaron Wisner invented "Automotive Electronic Cruise Control" in 1968 as an engineer for RCA's Industrial and Automation Systems Division in Plymouth, Michigan. His invention described in two patents filed that year (US 3570622 & US 3511329), with the second modifying his original design by debuting digital memory, was the first electronic device in controlling a car. Two decades passed before an integrated circuit for his design was developed by Motorola. as the MC14460 Auto Speed Control Processor in CMOS. The advantage of electronic speed control over its mechanical predecessor was that it could be integrated with electronic accident avoidance and engine management systems.

Following the 1973 oil crisis and rising fuel prices, the device became more popular in the U.S. "Cruise control can save gas by avoiding surges that expel fuel" while driving at steady speeds. In 1974, AMC, GM, and Chrysler priced the option at $60 to $70, while Ford charged $103.

Operation

The driver must bring the vehicle up to speed manually and use a button to set the cruise control to the current speed.

The cruise control takes its speed signal from a rotating driveshaft, speedometer cable, wheel speed sensor from the engine's RPM, or from internal speed pulses produced electronically by the vehicle. Most systems do not allow the use of the cruise control below a certain speed - typically around 25 mph (40 km/h). The vehicle will maintain the desired speed by pulling the throttle cable with a solenoid, a vacuum driven servomechanism, or by using the electronic systems built into the vehicle (fully electronic) if it uses a 'drive-by-wire' system.

All cruise control systems must be capable of being turned off both explicitly and automatically when the driver depresses the brake, and often also the clutch. Cruise control often includes a memory feature to resume the set speed after braking, and a coast feature to reduce the set speed without braking. When the cruise control is engaged, the throttle can still be used to accelerate the car, but once the pedal is released the car will then slow down until it reaches the previously set speed.

On the latest vehicles fitted with electronic throttle control, cruise control can be easily integrated into the vehicle's engine management system. Modern "adaptive" systems include the ability to automatically reduce speed when the distance to a car in front, or the speed limit, decreases. This is an advantage for those driving in unfamiliar areas.

The cruise control systems of some vehicles incorporate a "speed limiter" function, which will not allow the vehicle to accelerate beyond a pre-set maximum; this can usually be overridden by fully depressing the accelerator pedal. (Most systems will prevent the vehicle accelerating beyond the chosen speed, but will not apply the brakes in the event of overspeeding downhill.)

On vehicles with a manual transmission, cruise control is less flexible because the act of depressing the clutch pedal and shifting gears usually disengages the cruise control. The "resume" feature has to be used each time after selecting the new gear and releasing the clutch. Therefore, cruise control is of most benefit at motorway/highway speeds when top gear is used virtually all the time.

Advantages and Disadvantages

- Its usefulness for long drives (reducing driver fatigue, improving comfort by allowing positioning changes more safely) across highways and sparsely populated roads.

- Some drivers use it to avoid subconsciously violating speed limits. A driver who otherwise tends to subconsciously increase speed over the course of a highway journey may avoid speeding.

However, when used incorrectly cruise control can lead to accidents due to several factors, such as:

- speeding around curves that require slowing down

- rough or loose terrain that could negatively affect the cruise control controls

- rainy or wet weather could lose traction

Adaptive Cruise Control

Some modern vehicles have adaptive cruise control (ACC) systems, which is a general term meaning improved cruise control. These improvements can be automatic braking or dynamic set-speed type controls.

Automatic Braking Type: The automatic braking type use either a radar or laser set-up to allow the vehicle to keep pace with the car it is following, slow when closing in on the vehicle in front and accelerating again to the preset speed when traffic allows. Some systems also feature forward collision warning systems, which warns the driver if a vehicle in front—given the speed of both vehicles—gets too close (within the preset headway or braking distance).

Dynamic Set Speed Type: The dynamic set speed uses the GPS position of speed limit signs, from a database. Some are modifiable by the driver. At least one, Wikispeedia, incorporates crowdsourcing, so driver input is shared, improving the database for all users.

Non-Braking Type: The speed can be adjusted to allow traffic calming. One visual method uses OpenCV

References

- Steger, Carsten; Markus Ulrich & Christian Wiedemann (2008). Machine Vision Algorithms and Applications. Weinheim: Wiley-VCH. p. 1. ISBN 978-3-527-40734-7. Retrieved 2010-11-05.

- Graves, Mark & Bruce G. Batchelor (2003). Machine Vision for the Inspection of Natural Products. Springer. p. 5. ISBN 978-1-85233-525-0. Retrieved 2010-11-02.

- Relf, Christopher G. (2004). Image Acquisition and Processing with LabVIEW. 1. CRC Press. ISBN 978-0-8493-1480-3. Retrieved 2010-11-02.

- Hornberg, Alexander (2006). Handbook of Machine Vision. Wiley-VCH. p. 709. ISBN 978-3-527-40584-8. Retrieved 2010-11-05.

- Davies, E.R. (1996). Machine Vision - Theory Algorithms Practicalities (2nd ed.). Harcourt & Company. ISBN 978-0-12-206092-2.

- Demant C.; Streicher-Abel B. & Waszkewitz P. (1999). Industrial Image Processing: Visual Quality Control in Manufacturing. Springer-Verlag. p. 96. ISBN 3-540-66410-6.

- Leondes, Cornelius T. (2002). Expert systems: the technology of knowledge management and decision making for the 21st century. pp. 1–22. ISBN 978-0-12-443880-4.

- Russell, Stuart; Norvig, Peter (1995). Artificial Intelligence: A Modern Approach (PDF). Simon & Schuster. pp. 22–23. ISBN 0-13-103805-2. Retrieved 14 June 2014.

- Hurwitz, Judith (2011). Smart or Lucky: How Technology Leaders Turn Chance into Success. John Wiley & Son. p. 164. ISBN 1118033787. Retrieved 29 November 2013.

- Hayes-Roth, Frederick; Waterman, Donald; Lenat, Douglas (1983). Building Expert Systems. Addison-Wesley. ISBN 0-201-10686-8.

- Feigenbaum, Edward A.; McCorduck, Pamela (1983), The fifth generation (1st ed.), Reading, MA: Addison-Wesley, ISBN 978-0-201-11519-2, OCLC 9324691

- Thompson, C.; White, J.; Dougherty, B.; Schmidt, D. C. (2009). "Optimizing Mobile Application Performance with Model–Driven Engineering". Software Technologies for Embedded and Ubiquitous Systems (PDF). Lecture Notes in Computer Science. 5860. p. 36. doi:10.1007/978-3-642-10265-3_4. ISBN 978-3-642-10264-6.

- Leijdekkers, P. (2006). "Personal Heart Monitoring and Rehabilitation System using Smart Phones". 2006 International Conference on Mobile Business. p. 29. doi:10.1109/ICMB.2006.39. ISBN 0-7695-2595-4.

- A. W. Colombo, T. Bangemann, S. Karnouskos, J. Delsing, P. Stluka, R. Harrison, F. Jammes, and J. Lastra: Industrial Cloud-based Cyber- Physical Systems: The IMC-AESOP Approach. Springer Verlag, 2014, ISBN 978-3-319-05623-4.

- J .Fitzgerald, P.G. Larsen, M. Verhoef (Eds.): Collaborative Design for Embedded Systems: Co-modelling and Co-simulation. Springer Verlag, 2014, ISBN 978-3-642-54118-6.

- Dechow, David (February 2013). "Explore the Fundamentals of Machine Vision: Part 1". Vision Systems Design. 18 (2): 14–15. Retrieved 2013-03-05.

- Haskin, David (January 16, 2003). "Years After Hype, 'Expert Systems' Paying Off For Some". Datamation. Retrieved 29 November 2013.

- SAP News Desk. "SAP News Desk IntelliCorp Announces Participation in SAP EcoHub". laszlo.sys-con.com. LaszloTrack. Retrieved 29 November 2013.

- MacGregor, Robert (June 1991). "Using a description classifier to enhance knowledge representation". IEEE Expert. 6 (3): 41–46. doi:10.1109/64.87683. Retrieved 10 November 2013.

- Wong, Bo K.; Monaco, John A.; Monaco (September 1995). "Expert system applications in business: a review and analysis of the literature". Information and Management. 29 (3): 141–152. doi:10.1016/0378-7206(95)00023-p. Retrieved 29 November 2013.

Concepts and Principles of Robotics

Robotics is concerned with the design, construction and application of automated machines that can replace human beings in hazardous conditions or resemble humans in appearance. This chapter introduces the reader to concepts and principles like pneumatic cylinder, mobile service systems, parallel manipulator, articulated robot, robot-assisted surgery, bio-inspired robotics and delta robot. Mechatronics is best understood in confluence with the major topics listed in the following chapter.

Robotics

Robotics is the branch of mechanical engineering, electrical engineering and computer science that deals with the design, construction, operation, and application of robots, as well as computer systems for their control, sensory feedback, and information processing.

The Shadow robot hand system

These technologies deal with automated machines that can take the place of humans in dangerous environments or manufacturing processes, or resemble humans in appearance, behaviour, and or cognition. Many of today's robots are inspired by nature, contributing to the field of bio-inspired robotics.

The concept of creating machines that can operate autonomously dates back to classical times, but research into the functionality and potential uses of robots did not grow substantially until the 20th century. Throughout history, it has been frequently assumed that robots will one day be able to mimic human behavior and manage tasks in a human-like fashion. Today, robotics is a rapidly growing field, as technological advances continue; researching, designing, and building new robots serve various practical purposes, whether domestically, commercially, or militarily. Many robots are built to do jobs that are hazardous to people such as defusing bombs, finding survivors in unstable ruins, and exploring mines and shipwrecks. Robotics is also used in STEM (Science, Technology, Engineering, and Mathematics) as a teaching aid.

Etymology

The word *robotics* was derived from the word *robot*, which was introduced to the public by Czech writer Karel Čapek in his play *R.U.R. (Rossum's Universal Robots)*, which was published in 1920. The word *robot* comes from the Slavic word *robota*, which means labour. The play begins in a factory that makes artificial people called *robots*, creatures who can be mistaken for humans – very similar to the modern ideas of androids. Karel Čapek himself did not coin the word. He wrote a short letter in reference to an etymology in the *Oxford English Dictionary* in which he named his brother Josef Čapek as its actual originator.

According to the *Oxford English Dictionary*, the word *robotics* was first used in print by Isaac Asimov, in his science fiction short story "Liar!", published in May 1941 in *Astounding Science Fiction*. Asimov was unaware that he was coining the term; since the science and technology of electrical devices is *electronics*, he assumed *robotics* already referred to the science and technology of robots. In some of Asimov's other works, he states that the first use of the word *robotics* was in his short story *Runaround* (Astounding Science Fiction, March 1942). However, the original publication of "Liar!" predates that of "Runaround" by ten months, so the former is generally cited as the word's origin.

History of Robotics

In 1942 the science fiction writer Isaac Asimov created his Three Laws of Robotics.

In 1948 Norbert Wiener formulated the principles of cybernetics, the basis of practical robotics.

Fully autonomous robots only appeared in the second half of the 20th century. The first digitally operated and programmable robot, the Unimate, was installed in 1961 to lift hot pieces of metal from a die casting machine and stack them. Commercial and industrial robots are widespread today and used to perform jobs more cheaply, more accurately and more reliably, than humans. They are also employed in some jobs which are too dirty, dangerous, or dull to be suitable for humans. Robots are widely used in manufacturing, assembly, packing and packaging, transport, earth and space exploration, surgery, weaponry, laboratory research, safety, and the mass production of consumer and industrial goods.

Date	Significance	Robot Name	Inventor
Third century B.C. and earlier	One of the earliest descriptions of automata appears in the *Lie Zi* text, on a much earlier encounter between King Mu of Zhou (1023–957 BC) and a mechanical engineer known as Yan Shi, an 'artificer'. The latter allegedly presented the king with a life-size, human-shaped figure of his mechanical handiwork.		Yan Shi (Chinese: 偃师)
First century A.D. and earlier	Descriptions of more than 100 machines and automata, including a fire engine, a wind organ, a coin-operated machine, and a steam-powered engine, in *Pneumatica* and *Automata* by Heron of Alexandria		Ctesibius, Philo of Byzantium, Heron of Alexandria, and others
c. 420 B.C.E	A wooden, steam propelled bird, which was able to fly		Archytas of Tarentum
1206	Created early humanoid automata, programmable automaton band	Robot band, hand-washing automaton, automated moving peacocks	Al-Jazari
1495	Designs for a humanoid robot	Mechanical Knight	Leonardo da Vinci
1738	Mechanical duck that was able to eat, flap its wings, and excrete	Digesting Duck	Jacques de Vaucanson
1898	Nikola Tesla demonstrates first radio-controlled vessel.	Teleautomaton	Nikola Tesla
1921	First fictional automatons called "robots" appear in the play *R.U.R.*	Rossum's Universal Robots	Karel Čapek
1930s	Humanoid robot exhibited at the 1939 and 1940 World's Fairs	Elektro	Westinghouse Electric Corporation
1946	First general-purpose digital computer	Whirlwind	Multiple people
1948	Simple robots exhibiting biological behaviors	Elsie and Elmer	William Grey Walter
1956	First commercial robot, from the Unimation company founded by George Devol and Joseph Engelberger, based on Devol's patents	Unimate	George Devol
1961	First installed industrial robot.	Unimate	George Devol
1973	First industrial robot with six electromechanically driven axes	Famulus	KUKA Robot Group
1974	The world's first microcomputer controlled electric industrial robot, IRB 6 from ASEA, was delivered to a small mechanical engineering company in southern Sweden. The design of this robot had been patented already 1972.	IRB 6	ABB Robot Group
1975	Programmable universal manipulation arm, a Unimation product	PUMA	Victor Scheinman

Robotic Aspects

Robotic construction

Electrical aspect

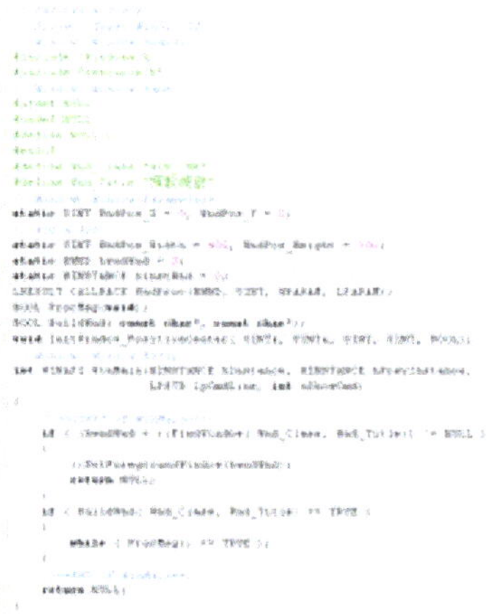

A level of programming

There are many types of robots; they are used in many different environments and for many different uses, although being very diverse in application and form they all share three basic similarities when it comes to their construction:

1. Robots all have some kind of mechanical construction, a frame, form or shape designed to achieve a particular task. For example, a robot designed to travel across heavy dirt or mud, might use caterpillar tracks. The mechanical aspect is mostly the creator's solution to completing the assigned task and dealing with the physics of the environment around it. Form follows function.

2. Robots have electrical components which power and control the machinery. For example, the robot with caterpillar tracks would need some kind of power to move the tracker treads. That power comes in the form of electricity, which will have to travel through a wire and originate from a battery, a basic electrical circuit. Even petrol powered machines that get their power mainly from petrol still require an electric current to start the combustion process which is why most petrol powered machines like cars, have batteries. The electrical aspect of robots is used for movement (through motors), sensing (where electrical signals are used to measure things like heat, sound, position, and energy status) and operation (robots need some level of electrical energy supplied to their motors and sensors in order to activate and perform basic operations)

3. All robots contain some level of computer programming code. A program is how a robot decides when or how to do something. In the caterpillar track example, a robot that needs to move across a muddy road may have the correct mechanical construction, and receive the correct amount of power from its battery, but would not go anywhere without a program telling it to move. Programs are the core essence of a robot, it could have excellent mechanical and electrical construction, but if its program is poorly constructed its performance will be very poor (or it may not perform at all). There are three different types of robotic programs: remote control, artificial intelligence and hybrid. A robot with remote control programing has a preexisting set of commands that it will only perform if and when it receives a signal from a control source, typically a human being with a remote control. It is perhaps more appropriate to view devices controlled primarily by human commands as falling in the discipline of automation rather than robotics. Robots that use artificial intelligence interact with their environment on their own without a control source, and can determine reactions to objects and problems they encounter using their preexisting programming. Hybrid is a form of programming that incorporates both AI and RC functions.

Applications

As more and more robots are designed for specific tasks this method of classification becomes more relevant. For example, many robots are designed for assembly work, which may not be readily adaptable for other applications. They are termed as "assembly robots". For seam welding, some suppliers provide complete welding systems with the robot i.e. the welding equipment along with other material handling facilities like turntables etc. as an integrated unit. Such an integrated robotic system is called a "welding robot" even though its discrete manipulator unit could be adapted to a variety of tasks. Some robots are specifically designed for heavy load manipulation, and are labelled as "heavy duty robots."

Current and potential applications include:

- Military robots

- Caterpillar plans to develop remote controlled machines and expects to develop fully autonomous heavy robots by 2021. Some cranes already are remote controlled.

- It was demonstrated that a robot can perform a herding task.

- Robots are increasingly used in manufacturing (since the 1960s). In the auto industry they can amount for more than half of the "labor". There are even "lights off" factories such as an IBM keyboard manufacturing factory in Texas that is 100% automated.

- Robots such as HOSPI are used as couriers in hospitals (hospital robot). Other hospital tasks performed by robots are receptionists, guides and porters helpers,

- Robots can serve as waiters and cooks., also at home. Boris is a robot that can load a dishwasher.

- Robot combat for sport – hobby or sport event where two or more robots fight in an arena to disable each other. This has developed from a hobby in the 1990s to several TV series worldwide.

- Cleanup of contaminated areas, such as toxic waste or nuclear facilities.

- Agricultural robots (AgRobots,).

- Domestic robots, cleaning and caring for the elderly

- Medical robots performing low-invasive surgery

- Household robots with full use.

- Nanorobots

Components

Power Source

At present mostly (lead–acid) batteries are used as a power source. Many different types of batteries can be used as a power source for robots. They range from lead–acid batteries, which are safe and have relatively long shelf lives but are rather heavy compared to silver–cadmium batteries that are much smaller in volume and are currently much more expensive. Designing a battery-powered robot needs to take into account factors such as safety, cycle lifetime and weight. Generators, often some type of internal combustion engine, can also be used. However, such designs are often mechanically complex and need fuel, require heat dissipation and are relatively heavy. A tether connecting the robot to a power supply would remove the power supply from the robot entirely. This has the advantage of saving weight and space by moving all power generation and storage components elsewhere. However, this design does come with the drawback of constantly having a cable connected to the robot, which can be difficult to manage. Potential power sources could be:

- pneumatic (compressed gases)

- Solar power (using the sun's energy and converting it into electrical power)

- hydraulics (liquids)

- flywheel energy storage

- organic garbage (through anaerobic digestion)

- nuclear

Actuation

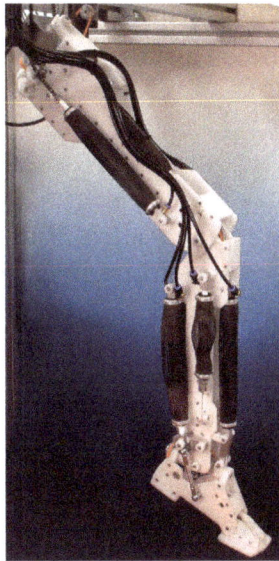

A robotic leg powered by air muscles

Actuators are the "muscles" of a robot, the parts which convert stored energy into movement. By far the most popular actuators are electric motors that rotate a wheel or gear, and linear actuators that control industrial robots in factories. There are some recent advances in alternative types of actuators, powered by electricity, chemicals, or compressed air.

Electric Motors

The vast majority of robots use electric motors, often brushed and brushless DC motors in portable robots or AC motors in industrial robots and CNC machines. These motors are often preferred in systems with lighter loads, and where the predominant form of motion is rotational.

Linear Actuators

Various types of linear actuators move in and out instead of by spinning, and often have quicker direction changes, particularly when very large forces are needed such as with industrial robotics. They are typically powered by compressed air (pneumatic actuator) or an oil (hydraulic actuator).

Series Elastic Actuators

A spring can be designed as part of the motor actuator, to allow improved force control. It has been used in various robots, particularly walking humanoid robots.

Air Muscles

Pneumatic artificial muscles, also known as air muscles, are special tubes that expand(-typically up to 40%) when air is forced inside them. They are used in some robot applications.

Muscle Wire

Muscle wire, also known as shape memory alloy, Nitinol® or Flexinol® wire, is a material which contracts (under 5%) when electricity is applied. They have been used for some small robot applications.

Electroactive Polymers

EAPs or EPAMs are a new plastic material that can contract substantially (up to 380% activation strain) from electricity, and have been used in facial muscles and arms of humanoid robots, and to enable new robots to float, fly, swim or walk.

Piezo Motors

Recent alternatives to DC motors are piezo motors or ultrasonic motors. These work on a fundamentally different principle, whereby tiny piezoceramic elements, vibrating many thousands of times per second, cause linear or rotary motion. There are different mechanisms of operation; one type uses the vibration of the piezo elements to step the motor in a circle or a straight line. Another type uses the piezo elements to cause a nut to vibrate or to drive a screw. The advantages of these motors are nanometer resolution, speed, and available force for their size. These motors are already available commercially, and being used on some robots.

Elastic Nanotubes

Elastic nanotubes are a promising artificial muscle technology in early-stage experimental development. The absence of defects in carbon nanotubes enables these filaments to deform elastically by several percent, with energy storage levels of perhaps 10 J/cm³ for metal nanotubes. Human biceps could be replaced with an 8 mm diameter wire of this material. Such compact "muscle" might allow future robots to outrun and outjump humans.

Sensing

Sensors allow robots to receive information about a certain measurement of the environment, or internal components. This is essential for robots to perform their tasks, and act upon any changes in the environment to calculate the appropriate response. They are used for various forms of measurements, to give the robots warnings about safety or malfunctions, and to provide real time information of the task it is performing.

Touch

Current robotic and prosthetic hands receive far less tactile information than the human hand. Recent research has developed a tactile sensor array that mimics the mechanical properties and touch receptors of human fingertips. The sensor array is constructed as a rigid core surrounded by conductive fluid contained by an elastomeric skin. Electrodes are mounted on the surface of the rigid core and are connected to an impedance-measuring device within the core. When the artificial skin touches an object the fluid path around the electrodes is deformed, producing impedance changes that map the forces received from the object. The researchers expect that an important function of such artificial fingertips will be adjusting robotic grip on held objects.

Scientists from several European countries and Israel developed a prosthetic hand in 2009, called SmartHand, which functions like a real one—allowing patients to write with it, type on a keyboard, play piano and perform other fine movements. The prosthesis has sensors which enable the patient to sense real feeling in its fingertips.

Vision

Computer vision is the science and technology of machines that see. As a scientific discipline, computer vision is concerned with the theory behind artificial systems that extract information from images. The image data can take many forms, such as video sequences and views from cameras.

In most practical computer vision applications, the computers are pre-programmed to solve a particular task, but methods based on learning are now becoming increasingly common.

Computer vision systems rely on image sensors which detect electromagnetic radiation which is typically in the form of either visible light or infra-red light. The sensors are designed using solid-state physics. The process by which light propagates and reflects off surfaces is explained using optics. Sophisticated image sensors even require quantum mechanics to provide a complete understanding of the image formation process. Robots can also be equipped with multiple vision sensors to be better able to compute the sense of depth in the environment. Like human eyes, robots' "eyes" must also be able to focus on a particular area of interest, and also adjust to variations in light intensities.

There is a subfield within computer vision where artificial systems are designed to mimic the processing and behavior of biological system, at different levels of complexity. Also, some of the learning-based methods developed within computer vision have their background in biology.

Other

Other common forms of sensing in robotics use lidar, radar and sonar.

Manipulation

KUKA industrial robot operating in a foundry

Baxter, a modern and versatile industrial robot developed by Rodney Brooks

Puma, one of the first industrial robots

Robots need to manipulate objects; pick up, modify, destroy, or otherwise have an effect. Thus the "hands" of a robot are often referred to as *end effectors*, while the "arm" is referred to as a *manipulator*. Most robot arms have replaceable effectors, each allowing them to perform some small range of tasks. Some have a fixed manipulator which cannot be replaced, while a few have one very general purpose manipulator, for example a humanoid hand. Learning how to manipulate a robot often requires a close feedback between human to the robot, although there are several methods for remote manipulation of robots.

Mechanical Grippers

One of the most common effectors is the gripper. In its simplest manifestation it consists of just two fingers which can open and close to pick up and let go of a range of small objects. Fingers can for example be made of a chain with a metal wire run through it. Hands that resemble and work more like a human hand include the Shadow Hand and the Robonaut hand. Hands that are of a mid-level complexity include the Delft hand. Mechanical grippers can come in various types, including friction and encompassing jaws. Friction jaws use all the force of the gripper to hold the object in place using friction. Encompassing jaws cradle the object in place, using less friction.

Vacuum Grippers

Vacuum grippers are very simple astrictive devices, but can hold very large loads provided the prehension surface is smooth enough to ensure suction.

Pick and place robots for electronic components and for large objects like car windscreens, often use very simple vacuum grippers.

General Purpose Effectors

Some advanced robots are beginning to use fully humanoid hands, like the Shadow Hand, MANUS, and the Schunk hand. These are highly dexterous manipulators, with as many as 20 degrees of freedom and hundreds of tactile sensors.

Locomotion

Rolling Robots

Segway in the Robot museum in Nagoya

For simplicity most mobile robots have four wheels or a number of continuous tracks. Some researchers have tried to create more complex wheeled robots with only one or two wheels. These can have certain advantages such as greater efficiency and reduced parts, as well as allowing a robot to navigate in confined places that a four-wheeled robot would not be able to.

Two-wheeled Balancing Robots

Balancing robots generally use a gyroscope to detect how much a robot is falling and then drive the wheels proportionally in the same direction, to counterbalance the fall at hundreds of times per second, based on the dynamics of an inverted pendulum. Many different balancing robots have been designed. While the Segway is not commonly thought of as a robot, it can be thought of as a component of a robot, when used as such Segway refer to them as RMP (Robotic Mobility Platform). An example of this use has been as NASA's Robonaut that has been mounted on a Segway.

One-wheeled Balancing Robots

A one-wheeled balancing robot is an extension of a two-wheeled balancing robot so that it can move in any 2D direction using a round ball as its only wheel. Several one-wheeled balancing robots have been designed recently, such as Carnegie Mellon University's "Ballbot" that is the approximate height and width of a person, and Tohoku Gakuin University's "BallIP". Because of the long, thin shape and ability to maneuver in tight spaces, they have the potential to function better than other robots in environments with people.

Spherical Orb Robots

Several attempts have been made in robots that are completely inside a spherical ball, either by spinning a weight inside the ball, or by rotating the outer shells of the sphere. These have also been referred to as an orb bot or a ball bot.

Six-wheeled Robots

Using six wheels instead of four wheels can give better traction or grip in outdoor terrain such as on rocky dirt or grass.

Tracked Robots

TALON military robots used by the United States Army

Tank tracks provide even more traction than a six-wheeled robot. Tracked wheels behave as if they were made of hundreds of wheels, therefore are very common for outdoor and military robots, where the robot must drive on very rough terrain. However, they are difficult to use indoors such as on carpets and smooth floors. Examples include NASA's Urban Robot "Urbie".

Walking Applied to Robots

Walking is a difficult and dynamic problem to solve. Several robots have been made which can walk reliably on two legs, however none have yet been made which are as robust as a human. There has been much study on human inspired walking, such as AMBER lab which was established in 2008 by the Mechanical Engineering Department at Texas A&M University. Many other robots have been built that walk on more than two legs, due to these robots being significantly easier to construct. Walking robots can be used for uneven terrains, which would provide better mobility and energy efficiency than other locomotion methods. Hybrids too have been proposed in movies such as I, Robot, where they walk on 2 legs and switch to 4 (arms+legs) when going to a sprint. Typically, robots on 2 legs can walk well on flat floors and can occasionally walk up stairs. None can walk over rocky, uneven terrain. Some of the methods which have been tried are:

ZMP Technique

The Zero Moment Point (ZMP) is the algorithm used by robots such as Honda's ASIMO. The robot's onboard computer tries to keep the total inertial forces (the combination of Earth's gravity and the acceleration and deceleration of walking), exactly opposed by the floor reaction force (the force of the floor pushing back on the robot's foot). In this way, the two forces cancel out, leaving no moment (force causing the robot to rotate and fall over). However, this is not exactly how a human walks, and the difference is obvious to human observers, some of whom have pointed out that ASIMO walks as if it needs the lavatory. ASIMO's walking algorithm is not static, and some dynamic balancing is used. However, it still requires a smooth surface to walk on.

Hopping

Several robots, built in the 1980s by Marc Raibert at the MIT Leg Laboratory, successfully demonstrated very dynamic walking. Initially, a robot with only one leg, and a very small foot, could stay upright simply by hopping. The movement is the same as that of a person on a pogo stick. As the robot falls to one side, it would jump slightly in that direction, in order to catch itself. Soon, the algorithm was generalised to two and four legs. A bipedal robot was demonstrated running and even performing somersaults. A quadruped was also demonstrated which could trot, run, pace, and bound. For a full list of these robots.

Dynamic Balancing (Controlled Falling)

A more advanced way for a robot to walk is by using a dynamic balancing algorithm, which is potentially more robust than the Zero Moment Point technique, as it constantly monitors the robot's motion, and places the feet in order to maintain stability. This technique was recently demonstrated by Anybots' Dexter Robot, which is so stable, it can even jump. Another example is the TU Delft Flame.

Passive Dynamics

Perhaps the most promising approach utilizes passive dynamics where the momentum of swinging limbs is used for greater efficiency. It has been shown that totally unpowered humanoid mechanisms can walk down a gentle slope, using only gravity to propel themselves. Using this technique, a robot need only supply a small amount of motor power to walk along a flat surface or a little more to walk up a hill. This technique promises to make walking robots at least ten times more efficient than ZMP walkers, like ASIMO.

Other Methods of Locomotion

Flying

Two robot snakes. Left one has 64 motors (with 2 degrees of freedom per segment), the right one 10.

A modern passenger airliner is essentially a flying robot, with two humans to manage it. The autopilot can control the plane for each stage of the journey, including takeoff, normal flight, and even landing. Other flying robots are uninhabited, and are known as unmanned aerial vehicles (UAVs). They can be smaller and lighter without a human pilot on board, and fly into dangerous territory for military surveillance missions. Some can even fire on targets under command. UAVs are also being developed which can fire on targets automatically, without the need for a command from a human. Other flying robots include cruise missiles, the Entomopter, and the Epson micro helicopter robot. Robots such as the Air Penguin, Air Ray, and Air Jelly have lighter-than-air bodies, propelled by paddles, and guided by sonar.

Snaking

Several snake robots have been successfully developed. Mimicking the way real snakes move, these robots can navigate very confined spaces, meaning they may one day be used to search for people trapped in collapsed buildings. The Japanese ACM-R5 snake robot can even navigate both on land and in water.

Skating

A small number of skating robots have been developed, one of which is a multi-mode walking and skating device. It has four legs, with unpowered wheels, which can either step or roll. Another robot, Plen, can use a miniature skateboard or roller-skates, and skate across a desktop.

Capuchin, a climbing robot

Climbing

Several different approaches have been used to develop robots that have the ability to climb vertical surfaces. One approach mimics the movements of a human climber on a wall with protrusions; adjusting the center of mass and moving each limb in turn to gain leverage. An example of this is Capuchin, built by Dr. Ruixiang Zhang at Stanford University, California. Another approach uses the specialized toe pad method of wall-climbing geckoes, which can run on smooth surfaces such as vertical glass. Examples of this approach include Wallbot and Stickybot. China's *Technology Daily* reported on November 15, 2008 that Dr. Li Hiu Yeung and his research group of New Concept Aircraft (Zhuhai) Co., Ltd. had successfully developed a bionic gecko robot named "Speedy Freelander". According to Dr. Li, the gecko robot could rapidly climb up and down a variety of building walls, navigate through ground and wall fissures, and walk upside-down on the ceiling. It was also able to adapt to the surfaces of smooth glass, rough, sticky or dusty walls as well as various types of metallic materials. It could also identify and circumvent obstacles automatically. Its flexibility and speed were comparable to a natural gecko. A third approach is to mimic the motion of a snake climbing a pole.. Lastly one may mimic the movements of a human climber on a wall with protrusions; adjusting the center of mass and moving each limb in turn to gain leverage.

Swimming (Piscine)

It is calculated that when swimming some fish can achieve a propulsive efficiency greater than 90%. Furthermore, they can accelerate and maneuver far better than any manmade boat or submarine, and produce less noise and water disturbance. Therefore, many researchers studying underwater robots would like to copy this type of locomotion. Notable examples are the Essex University Computer Science Robotic Fish G9, and the Robot Tuna built by the Institute of Field Robotics, to analyze and mathematically model thunniform motion. The Aqua Penguin, designed and built by Festo of Germany, copies the streamlined shape and propulsion by front "flippers" of penguins. Festo have also built the Aqua Ray and Aqua Jelly, which emulate the locomotion of manta ray, and jellyfish, respectively.

Robotic Fish: *iSplash*-II

In 2014 *iSplash*-II was developed by R.J Clapham PhD at Essex University. It was the first robotic fish capable of outperforming real carangiform fish in terms of average maximum velocity (measured in body lengths/ second) and endurance, the duration that top speed is maintained. This build attained swimming speeds of 11.6BL/s (i.e. 3.7 m/s). The first build, *iSplash*-I (2014) was the first robotic platform to apply a full-body length carangiform swimming motion which was found to increase swimming speed by 27% over the traditional approach of a posterior confined wave form.

Sailing

The autonomous sailboat robot *Vaimos*

Sailboat robots have also been developed in order to make measurements at the surface of the ocean. A typical sailboat robot is *Vaimos* built by IFREMER and ENSTA-Bretagne. Since the propulsion of sailboat robots uses the wind, the energy of the batteries is only used for the computer, for the communication and for the actuators (to tune the rudder and the sail). If the robot is equipped with solar panels, the robot could theoretically navigate forever. The two main competitions of sailboat robots are WRSC, which takes place every year in Europe, and Sailbot.

Environmental Interaction and Navigation

Radar, GPS, and lidar, are all combined to provide proper navigation and obstacle avoidance (vehicle developed for 2007 DARPA Urban Challenge)

Though a significant percentage of robots in commission today are either human controlled, or operate in a static environment, there is an increasing interest in robots that can operate autonomously in a dynamic environment. These robots require some combination of navigation hardware and software in order to traverse their environment. In particular unforeseen events (e.g. people and other obstacles that are not stationary) can cause problems or collisions. Some highly advanced robots such as ASIMO, and Meinü robot have particularly good robot navigation hardware and software. Also, self-controlled cars, Ernst Dickmanns' driverless car, and the entries in the DARPA Grand Challenge, are capable of sensing the environment well and subsequently making navigational decisions based on this information. Most of these robots employ a GPS navigation device with waypoints, along with radar, sometimes combined with other sensory data such as lidar, video cameras, and inertial guidance systems for better navigation between waypoints.

Human-robot Interaction

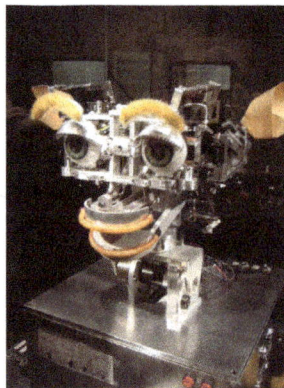

Kismet can produce a range of facial expressions.

The state of the art in sensory intelligence for robots will have to progress through several orders of magnitude if we want the robots working in our homes to go beyond vacuum-cleaning the floors. If robots are to work effectively in homes and other non-industrial environments, the way they are instructed to perform their jobs, and especially how they will be told to stop will be of critical importance. The people who interact with them may have little or no training in robotics, and so any interface will need to be extremely intuitive. Science fiction authors also typically assume that robots will eventually be capable of communicating with humans through speech, gestures, and facial expressions, rather than a command-line interface. Although speech would be the most natural way for the human to communicate, it is unnatural for the robot. It will probably be a long time before robots interact as naturally as the fictional C-3PO, or Data of Star Trek, Next Generation.

Speech Recognition

Interpreting the continuous flow of sounds coming from a human, in real time, is a difficult task for a computer, mostly because of the great variability of speech. The same word, spoken by the same person may sound different depending on local acoustics, volume, the previous word, whether or not the speaker has a cold, etc.. It becomes even harder when the speaker has a different accent. Nevertheless, great strides have been made in the field since Davis, Biddulph, and Balashek designed the first "voice input system" which recognized "ten digits spoken by a single user with 100% accuracy" in 1952. Currently, the best systems can recognize continuous, natural speech, up to 160 words per minute, with an accuracy of 95%.

Robotic Voice

Other hurdles exist when allowing the robot to use voice for interacting with humans. For social reasons, synthetic voice proves suboptimal as a communication medium, making it necessary to develop the emotional component of robotic voice through various techniques.

Gestures

One can imagine, in the future, explaining to a robot chef how to make a pastry, or asking directions from a robot police officer. In both of these cases, making hand gestures would aid the verbal descriptions. In the first case, the robot would be recognizing gestures made by the human, and perhaps repeating them for confirmation. In the second case, the robot police officer would gesture to indicate "down the road, then turn right". It is likely that gestures will make up a part of the interaction between humans and robots. A great many systems have been developed to recognize human hand gestures.

Facial Expression

Facial expressions can provide rapid feedback on the progress of a dialog between two humans, and soon may be able to do the same for humans and robots. Robotic faces have been constructed by Hanson Robotics using their elastic polymer called Frubber, allowing a large number of facial expressions due to the elasticity of the rubber facial coating and embedded subsurface motors (servos). The coating and servos are built on a metal skull. A robot should know how to approach a human, judging by their facial expression and body language. Whether the person is happy, frightened, or crazy-looking affects the type of interaction expected of the robot. Likewise, robots like Kismet and the more recent addition, Nexi can produce a range of facial expressions, allowing it to have meaningful social exchanges with humans.

Artificial Emotions

Artificial emotions can also be generated, composed of a sequence of facial expressions and/or gestures. As can be seen from the movie Final Fantasy: The Spirits Within, the programming of these artificial emotions is complex and requires a large amount of human observation. To simplify this programming in the movie, presets were created together with a special software program. This decreased the amount of time needed to make the film. These presets could possibly be transferred for use in real-life robots.

Personality

Many of the robots of science fiction have a personality, something which may or may not be desirable in the commercial robots of the future. Nevertheless, researchers are trying to create robots which appear to have a personality: i.e. they use sounds, facial expressions, and body language to try to convey an internal state, which may be joy, sadness, or fear. One commercial example is Pleo, a toy robot dinosaur, which can exhibit several apparent emotions.

Social Intelligence

The Socially Intelligent Machines Lab of the Georgia Institute of Technology researches new concepts of guided teaching interaction with robots. Aim of the projects is a social robot learns task goals from human demonstrations without prior knowledge of high-level concepts. These new concepts are grounded from low-level continuous sensor data through unsupervised learning, and task goals are subsequently learned using a Bayesian approach. These concepts can be used to transfer knowledge to future tasks, resulting in faster learning of those tasks. The results are demonstrated by the robot *Curi* who can scoop some pasta from a pot onto a plate and serve the sauce on top.

Control

The mechanical structure of a robot must be controlled to perform tasks. The control of a robot involves three distinct phases – perception, processing, and action (robotic paradigms). Sensors give information about the environment or the robot itself (e.g.

the position of its joints or its end effector). This information is then processed to be stored or transmitted, and to calculate the appropriate signals to the actuators (motors) which move the mechanical.

Puppet Magnus, a robot-manipulated marionette with complex control systems

RuBot II can resolve manually Rubik cubes

The processing phase can range in complexity. At a reactive level, it may translate raw sensor information directly into actuator commands. Sensor fusion may first be used to estimate parameters of interest (e.g. the position of the robot's gripper) from noisy sensor data. An immediate task (such as moving the gripper in a certain direction) is inferred from these estimates. Techniques from control theory convert the task into commands that drive the actuators.

At longer time scales or with more sophisticated tasks, the robot may need to build and reason with a "cognitive" model. Cognitive models try to represent the robot, the world, and how they interact. Pattern recognition and computer vision can be used to track objects. Mapping techniques can be used to build maps of the world. Finally, motion planning and other artificial intelligence techniques may be used to figure out how to act. For example, a planner may figure out how to achieve a task without hitting obstacles, falling over, etc.

Autonomy Levels

TOPIO, a humanoid robot, played ping pong at Tokyo IREX 2009.

Control systems may also have varying levels of autonomy.

1. Direct interaction is used for haptic or tele-operated devices, and the human has nearly complete control over the robot's motion.

2. Operator-assist modes have the operator commanding medium-to-high-level tasks, with the robot automatically figuring out how to achieve them.

3. An autonomous robot may go for extended periods of time without human interaction. Higher levels of autonomy do not necessarily require more complex cognitive capabilities. For example, robots in assembly plants are completely autonomous, but operate in a fixed pattern.

Another classification takes into account the interaction between human control and the machine motions.

1. Teleoperation. A human controls each movement, each machine actuator change is specified by the operator.

2. Supervisory. A human specifies general moves or position changes and the machine decides specific movements of its actuators.

3. Task-level autonomy. The operator specifies only the task and the robot manages itself to complete it.

4. Full autonomy. The machine will create and complete all its tasks without human interaction.

Robotics Research

Much of the research in robotics focuses not on specific industrial tasks, but on investigations into new types of robots, alternative ways to think about or design robots, and new ways to manufacture them but other investigations, such as MIT's cyberflora project, are almost wholly academic.

A first particular new innovation in robot design is the opensourcing of robot-projects. To describe the level of advancement of a robot, the term "Generation Robots" can be used. This term is coined by Professor Hans Moravec, Principal Research Scientist at the Carnegie Mellon University Robotics Institute in describing the near future evolution of robot technology. *First generation* robots, Moravec predicted in 1997, should have an intellectual capacity comparable to perhaps a lizard and should become available by 2010. Because the *first generation* robot would be incapable of learning, however, Moravec predicts that the *second generation* robot would be an improvement over the *first* and become available by 2020, with the intelligence maybe comparable to that of a mouse. The *third generation* robot should have the intelligence comparable to that of a monkey. Though *fourth generation* robots, robots with human intelligence, professor Moravec predicts, would become possible, he does not predict this happening before around 2040 or 2050.

The second is evolutionary robots. This is a methodology that uses evolutionary computation to help design robots, especially the body form, or motion and behavior controllers. In a similar way to natural evolution, a large population of robots is allowed to compete in some way, or their ability to perform a task is measured using a fitness function. Those that perform worst are removed from the population, and replaced by a new set, which have new behaviors based on those of the winners. Over time the population improves, and eventually a satisfactory robot may appear. This happens without any direct programming of the robots by the researchers. Researchers use this method both to create better robots, and to explore the nature of evolution. Because the process often requires many generations of robots to be simulated, this technique may be run entirely or mostly in simulation, then tested on real robots once the evolved algorithms are good enough. Currently, there are about 10 million industrial robots toiling around the world, and Japan is the top country having high density of utilizing robots in its manufacturing industry.

Dynamics and Kinematics

The study of motion can be divided into kinematics and dynamics. Direct kinematics refers to the calculation of end effector position, orientation, velocity, and acceleration when the corresponding joint values are known. Inverse kinematics refers to the opposite case in which required joint values are calculated for given end effector values, as done in path planning. Some special aspects of kinematics include handling of redundancy (different possibilities of performing the same movement), collision avoidance, and singularity avoidance. Once all relevant positions, velocities, and accelerations have been calculated using kinematics, methods from the field of dynamics are used to study the effect of forces upon these movements. Direct dynamics refers to the calculation of accelerations in the robot once the applied forces are known. Direct dynamics is used in computer simulations of the robot. Inverse dynamics refers to the calculation of the actuator forces necessary to create a prescribed end effector acceleration. This information can be used to improve the control algorithms of a robot.

In each area mentioned above, researchers strive to develop new concepts and strategies, improve existing ones, and improve the interaction between these areas. To do this, criteria for "optimal" performance and ways to optimize design, structure, and control of robots must be developed and implemented.

Bionics and Biomimetics

Bionics and biomimetics apply the physiology and methods of locomotion of animals to the design of robots. For example, the design of BionicKangaroo was based on the way kangaroos jump.

Education and Training

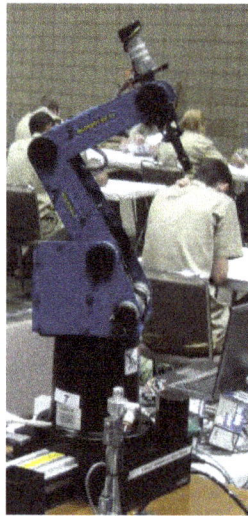

The SCORBOT-ER 4u educational robot

Robotics engineers design robots, maintain them, develop new applications for them, and conduct research to expand the potential of robotics. Robots have become a popular educational tool in some middle and high schools, particularly in parts of the USA, as well as in numerous youth summer camps, raising interest in programming, artificial intelligence and robotics among students. First-year computer science courses at some universities now include programming of a robot in addition to traditional software engineering-based coursework.

Career Training

Universities offer bachelors, masters, and doctoral degrees in the field of robotics. Vocational schools offer robotics training aimed at careers in robotics.

Certification

The Robotics Certification Standards Alliance (RCSA) is an international robotics certification authority that confers various industry- and educational-related robotics certifications.

Summer Robotics Camp

Several national summer camp programs include robotics as part of their core curriculum. In addition, youth summer robotics programs are frequently offered by celebrated museums such as the American Museum of Natural History and The Tech Museum of Innovation in Silicon Valley, CA, just to name a few.

Robotics Competitions

There are lots of competitions all around the globe. One of the most important competitions is the FLL or FIRST Lego League. The idea of this specific competition is that kids start developing knowledge and getting into robotics while playing with Legos since they are 9 years old. This competition is associated with Ni or National Instruments.

Robotics Afterschool Programs

Many schools across the country are beginning to add robotics programs to their after school curriculum. Some major programs for afterschool robotics include FIRST Robotics Competition, Botball and B.E.S.T. Robotics. Robotics competitions often include aspects of business and marketing as well as engineering and design.

The Lego company began a program for children to learn and get excited about robotics at a young age.

Employment

A robot technician builds small all-terrain robots. (Courtesy: MobileRobots Inc)

Robotics is an essential component in many modern manufacturing environments. As factories increase their use of robots, the number of robotics–related jobs grow and have been observed to be steadily rising. The employment of robots in industries has increased productivity and efficiency savings and is typically seen as a long term investment for benefactors.

Occupational Safety and Health Implications of Robotics

A discussion paper drawn up by EU-OSHA highlights how the spread of robotics presents both opportunities and challenges for occupational safety and health (OSH).

The greatest OSH benefits stemming from the wider use of robotics should be substitution for people working in unhealthy or dangerous environments. In space, defence, security, or the nuclear industry, but also in logistics, maintenance and inspection, autonomous robots are particularly useful in replacing human workers performing dirty, dull or unsafe tasks, thus avoiding workers' exposures to hazardous agents and conditions and reducing physical, ergonomic and psychosocial risks. For example, robots are already used to perform repetitive and monotonous tasks, to handle radioactive material or to work in explosive atmospheres. In the future, many other highly repetitive, risky or unpleasant tasks will be performed by robots in a variety of sectors like agriculture, construction, transport, healthcare, firefighting or cleaning services.

Despite these advances, there are certain skills to which humans will be better suited than machines for some time to come and the question is how to achieve the best combination of human and robot skills. The advantages of robotics include heavy-duty jobs with precision and repeatability, whereas the advantages of humans include creativity, decision-making, flexibility and adaptability. This need to combine optimal skills has resulted in collaborative robots and humans sharing a common workspace more closely and led to the development of new approaches and standards to guarantee the safety of the "man-robot merger". Some European countries are including robotics in their national programmes and trying to promote a safe and flexible co-operation between robots and operators to achieve better productivity. For example, the German Federal Institute for Occupational Safety and Health (BAuA) organises annual workshops on the topic "human-robot collaboration".

In future, co-operation between robots and humans will be diversified, with robots increasing their autonomy and human-robot collaboration reaching completely new forms. Current approaches and technical standards aiming to protect employees from the risk of working with collaborative robots will have to be revised.

Bio-inspired Robotics

Bio-inspired robotic locomotion is a fairly new subcategory of bio-inspired design. It is about learning concepts from nature and applying them to the design of real-world engineered systems. More specifically, this field is about making robots that are inspired by biological systems. Biomimicry and bio-inspired design are sometimes confused. Biomimicry is copying the nature while bio-inspired design is learning from nature and making a mechanism that is simpler and more effective than the system observed in nature. Biomimicry has led to the development of a different branch of robotics called soft robotics. The biological systems have been optimized for specific tasks according to their habitat. However, they are multifunctional and are not designed for only one specific functionality. Bio-inspired robotics is about studying biological systems, and look for the mechanisms that may solve a problem in the engineering field. The designer should then try to simplify and enhance that mechanism for the specific task of interest. Bio-inspired roboticists are usually interested in biosensors (e.g. eye), bioactuators (e.g. muscle), or biomaterials (e.g. spider silk). Most of the robots have some type of locomotion system. Thus, in this article different modes of animal locomotion and few examples of the corresponding bio-inspired robots are introduced.

Two u-CAT robots that are being developed at the Tallinn University of Technology to reduce the cost of underwater archaeological operations

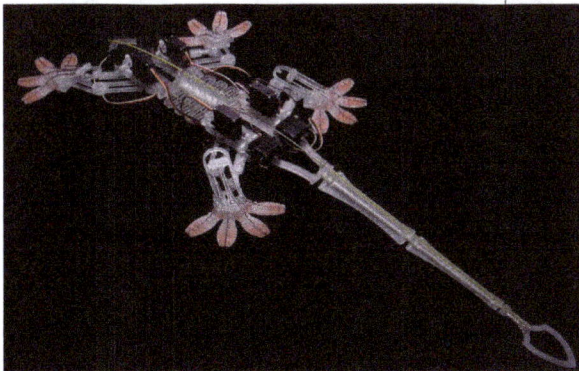

Stickybot: a gecko-inspired robot

Biolocomotion

Biolocomotion or animal locomotion is usually categorized as below:

Locomotion on a Surface

Locomotion on a surface may include terrestrial locomotion and arboreal locomotion. We will specifically discuss about terrestrial locomotion in detail in the next section.

Big eared townsend bat (*Corynorhinus townsendii*)

Locomotion in a Fluid

Locomotion in a blood stream swimming and flying. There are many swimming and flying robots designed and built by roboticists.

Behavioral Classification (Terrestrial Locomotion)

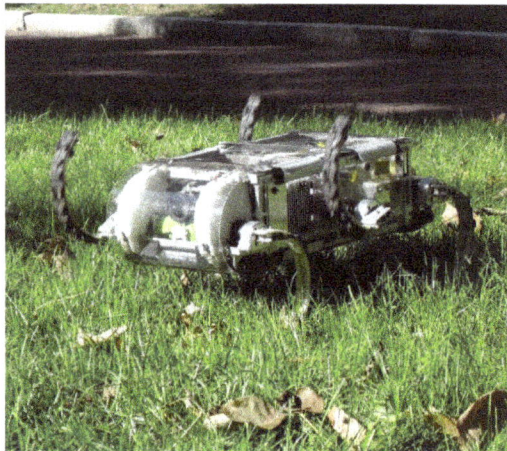

Rhex: a Reliable Hexapedal Robot

There are many animal and insects moving on land with or without legs. We will discuss about legged and limbless locomotion in this section as well as climbing and jumping. Anchoring the feet is fundamental to locomotion on land. The ability to increase traction is important for slip-free motion on surfaces such as smooth rock faces and ice, and is especially critical for moving uphill. Numerous biological mechanisms exist for providing purchase: claws rely upon friction-based mechanisms; gecko feet upon van der walls forces; and some insect feet upon fluid-mediated adhesive forces.

Legged Locomotion

Legged robots may have one, two, four, six, or many legs depending on the application. One of the main advantages of using legs instead of wheels is moving on uneven environment more effectively. Bipedal, quadrupedal, and hexapedal locomotion are among the most favorite types of legged locomotion in the field of bio-inspired robotics. Rhex, a Reliable Hexapedal robot and Cheetah are the two fastest running robots so far. iSprawl is another hexapedal robot inspired by cockroach locomotion that has been developed at Stanford University. This robot can run up to 15 body length per second and can achieve speeds of up to 2.3 m/s. The original version of this robot was pneumatically driven while the new generation uses a single electric motor for locomotion.

Limbless Locomotion

Terrain involving topography over a range of length scales can be challenging for most organisms and biomimetic robots. Such terrain are easily passed over by limbless organisms such as snakes. Several animals and insects including worms, snails, caterpillars, and snakes are capable of limbless locomotion. A review of snake-like robots is presented by Hirose et al. These robots can be categorized as robots with passive or active wheels, robots with active treads, and undulating robots using vertical waves or linear expansions. Most snake-like robots use wheels, which provide a forward-transverse frictional anisotropy. The majority of snake-like robots use either lateral undulation or rectilinear locomotion and have difficulty climbing vertically. Choset has recently developed a modular robot that can mimic several snake gaits, but it cannot perform concertina motion. Researchers at Georgia Tech have recently developed two snake-like robots called Scalybot. The focus of these robots is on the role of snake ventral scales on adjusting the frictional properties in different directions. These robots can actively control their scales to modify their frictional properties and move on a variety of surfaces efficiently.

Climbing

Climbing is an especially difficult task because mistakes made by the climber may cause the climber to lose its grip and fall. Most robots have been built around a single functionality observed in their biological counterparts. Geckobots typically use van der waals forces that work only on smooth surfaces. Stickybots, and use directional dry adhesives that works best on smooth surfaces. Spinybot and the RiSE robot are among the insect-like robots that use spines instead. Legged climbing robots have several limitations. They cannot handle large obstacles since they are not flexible and they require a wide space for moving. They usually cannot climb both smooth and rough surfaces or handle vertical to horizontal transitions as well.

Jumping

One of the tasks commonly performed by a variety of living organisms is jumping. Bharal, hares, kangaroo, grasshopper, flea, and locust are among the best jumping animals. A miniature 7g jumping robot inspired by locust has been developed at EPFL that can jump up to 138 cm. The jump event is induced by releasing the tension of a spring. ETH Zurich has reported a soft jumping robot based on the combustion of methane and laughing gas. The thermal gas expansion inside the soft combustion chamber drastically increases the chamber volume. This causes the 2 kg robot to jump up to 20 cm. The soft robot inspired by a roly-poly toy then reorientates itself into an upright position after landing.

Morphological Classification

Modular

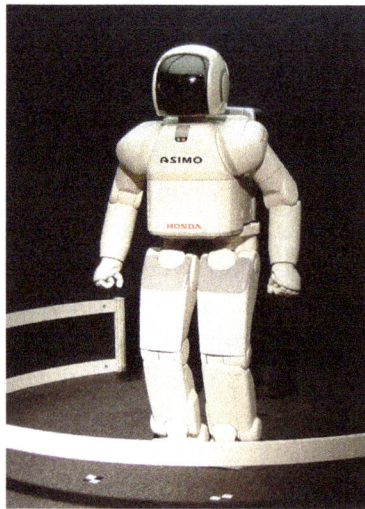

Honda Asimo: A Humanoid robot

The modular robots are typically capable of performing several tasks and are specifically useful for search and rescue or exploratory missions. Some of the featured robots in this category include a salamander inspired robot developed at EPFL that can walk and swim, a snake inspired robot developed at Carnegie-Mellon University that has four different modes of terrestrial locomotion, and a cockroach inspired robot can run and climb on a variety of complex terrain.

Humanoid

Humanoid robots are robots that look human-like or are inspired by the human form.

There are many different types of humanoid robots for applications such as personal assistance, reception, work at industries, or companionship. These type of robots are used for research purposes as well and were originally developed to build better orthosis and prosthesis for human beings. Petman is one of the first and most advanced humanoid robots developed at Boston Dynamics. Some of the humanoid robots such as Honda Asimo are over actuated. On the other hand, there are some humanoid robots like the robot developed at Cornell University that do not have any actuators and walk passively descending a shallow slope.

Swarming

The collective behavior of animals has been of interest to researchers for several years. Ants can make structures like rafts to survive on the rivers. Fish can sense their environment more effectively in large groups. Swarm robotics is a fairly new field and the goal is to make robots that can work together and transfer the data, make structures as a group, etc.

Soft

Soft robots are robots composed entirely of soft materials and moved through pneumatic pressure, similar to an octopus or starfish. Such robots are flexible enough to move in very limited spaces (such as in the human body). The first multigait soft robots was developed in 2011 and the first fully integrated, independent soft robot (with soft batteries and control systems) was developed in 2015

Articulated Robot

A six-axis articulated welding robot reaching into a fixture to weld.

An articulated robot is a robot with rotary joints (e.g. a legged robot or an industrial robot). Articulated robots can range from simple two-jointed structures to systems with 10 or more interacting joints. They are powered by a variety of means, including electric motors.

Some types of robots, such as robotic arms, can be articulated or non-articulated.

Articulated Robots in Action

Robots palletizing food (Bakery)

Manufacturing of steel bridges, cutting steel

Definitions

An articulated robot is one which uses rotary joints to access its work space. Usually the joints are arranged in a "chain", so that one joint supports another further in the chain.

Continuous Path: A control scheme whereby the inputs or commands specify every point along a desired path of motion. The path is controlled by the coordinated motion of the manipulator joints.

Degrees Of Freedom (DOF): The number of independent motions in which the end effector can move, defined by the number of axes of motion of the manipulator.

Gripper: A device for grasping or holding, attached to the free end of the last manipulator link; also called the robot's hand or end-effector.

Payload: The maximum payload is the amount of weight carried by the robot manipulator at reduced speed while maintaining rated precision. Nominal payload is measured at maximum speed while maintaining rated precision. These ratings are highly dependent on the size and shape of the payload.

Pick and place Cycle: Pick and place Cycle is the time, in seconds, to execute the following motion sequence: Move down one inch, grasp a rated payload; move up one inch; move across twelve inches; move down one inch; ungrasp; move up one inch; and return to start location.

Reach: The maximum horizontal distance from the center of the robot base to the end of its wrist.

Accuracy: The difference between the point that a robot is trying to achieve and the actual resultant position. Absolute accuracy is the difference between a point instructed by the robot control system and the point actually achieved by the manipulator arm, while repeatability is the cycle-to-cycle variation of the manipulator arm when aimed at the same point.

Repeatability: The ability of a system or mechanism to repeat the same motion or achieve the same points when presented with the same control signals. The cycle-to-cycle error of a system when trying to perform a specific task

Resolution: The smallest increment of motion or distance that can be de-tected or controlled by the control system of a mechanism. The resolution of any joint is a function of encoder pulses per revolution and drive ratio, and dependent on the distance between the tool center point and the joint axis.

Robot Program: A robot communication program for IBM and compatible personal computers. Provides terminal emulation and utility functions. This program can record all of the user memory, and some of the system memory to disk files.

Maximum Speed: The compounded maximum speed of the tip of a robot moving at full extension with all joints moving simultaneously in complimentary directions. This speed is the theoretical maximum and should under no circumstances be used to estimate cycle time for a particular application. A better measure of real world speed is the standard twelve inch pick and place cycle time. For critical applications, the best indicator of achievable cycle time is a physical simulation.

Servo Controlled: Controlled by a driving signal which is determined by the error between the mechanism's present position and the desired output position.

Via Point: A point through which the robot's tool should pass without stopping; via points are programmed in order to move beyond obstacles or to bring the arm into a lower inertia posture for part of the motion.

Work Envelope: A three-dimensional shape that defines the boundaries that the robot manipulator can reach; also known as reach envelope.

Robot-assisted Surgery

Robotic surgery, computer-assisted surgery, and robotically-assisted surgery are terms for technological developments that use robotic systems to aid in surgical procedures. Robotically-assisted surgery was developed to overcome the limitations of pre-existing minimally-invasive surgical procedures and to enhance the capabilities of surgeons performing open surgery.

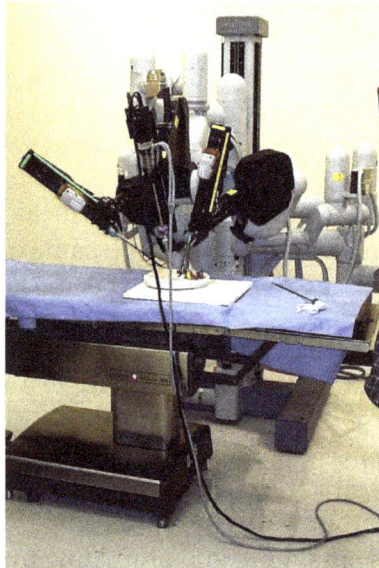

A robotically assisted surgical system used for prostatectomies, cardiac valve repair and gynecologic surgical procedures

In the case of robotically-assisted minimally-invasive surgery, instead of directly moving the instruments, the surgeon uses one of two methods to control the instruments; either a direct telemanipulator or through computer control. A telemanipulator is a remote manipulator that allows the surgeon to perform the normal movements associated with the surgery whilst the robotic arms carry out those movements using end-effectors and manipulators to perform the actual surgery on the patient. In computer-controlled systems the surgeon uses a computer to control the robotic arms and its end-effectors, though these systems can also still use telemanipulators for their input. One advantage of using the computerised method is that the surgeon does not have to be present, but can be anywhere in the world, leading to the possibility for remote surgery.

In the case of enhanced open surgery, autonomous instruments (in familiar configurations) replace traditional steel tools, performing certain actions (such as rib spreading) with much smoother, feedback-controlled motions than could be achieved by a human hand. The main object of such smart instruments is to reduce or eliminate the tissue trauma traditionally associated with open surgery without requiring more than a few minutes' training on the part of surgeons. This approach seeks to improve open surgeries, particularly cardio-thoracic, that have so far not benefited from minimally-invasive techniques.

Robotic surgery has been criticized for its expense, by one estimate costing $1,500 to $2000 more per patient.

Comparison to Traditional Methods

Major advances aided by surgical robots have been remote surgery, minimally invasive surgery and unmanned surgery. Due to robotic use, the surgery is done with precision, miniaturization, smaller incisions; decreased blood loss, less pain, and quicker healing time. Articulation beyond normal manipulation and three-dimensional magnification helps resulting in improved ergonomics. Due to these techniques there is a reduced duration of hospital stays, blood loss, transfusions, and use of pain medication. The existing open surgery technique has many flaws like limited access to surgical area, long recovery time, long hours of operation, blood loss, surgical scars and marks.

The robot normally costs $1,390,000 and while its disposable supply cost is normally $1,500 per procedure, the cost of the procedure is higher. Additional surgical training is needed to operate the system. Numerous feasibility studies have been done to determine whether the purchase of such systems are worthwhile. As it stands, opinions differ dramatically. Surgeons report that, although the manufacturers of such systems provide training on this new technology, the learning phase is intensive and surgeons must operate on twelve to eighteen patients before they adapt. During the training phase, minimally invasive operations can take up to twice as long as traditional surgery, leading to operating room tie ups and surgical staffs keeping patients under anesthesia for longer periods. Patient surveys indicate they chose the procedure based on expectations of decreased morbidity, improved outcomes, reduced blood loss and less pain. Higher expectations may explain higher rates of dissatisfaction and regret.

Compared with other minimally invasive surgery approaches, robot-assisted surgery gives the surgeon better control over the surgical instruments and a better view of the surgical site. In addition, surgeons no longer have to stand throughout the surgery and do not tire as quickly. Naturally occurring hand tremors are filtered out by the robot's computer software. Finally, the surgical robot can continuously be used by rotating surgery teams.

Critics of the system, including the American Congress of Obstetricians and Gynecologists, say there is a steep learning curve for surgeons who adopt use of the system and that there's a lack of studies that indicate long-term results are superior to results following traditional laparoscopic surgery.

A Medicare study found that some procedures that have traditionally been performed with large incisions can be converted to "minimally invasive" endoscopic procedures with the use of the Da Vinci Surgical System, shortening length-of-stay in the hospital and reducing recovery times. But because of the hefty cost of the robotic system it is not clear that it is cost-effective for hospitals and physicians despite any benefits to patients since there is no additional reimbursement paid by the government or insurance companies when the system is used.

Robot-assisted pancreatectomies have been found to be associated with "longer operating time, lower estimated blood loss, a higher spleen-preservation rate, and shorter hospital stay[s]" than laparoscopic pancreatectomies; there was "no significant difference in transfusion, conversion to open surgery, overall complications, severe complications, pancreatic fistula, severe pancreatic fistula, ICU stay, total cost, and 30-day mortality between the two groups."

Uses

General Surgery

In early 2000 the field of general surgical interventions with the daVinci device was explored by surgeons at Ohio State University. Reports were published in esophageal and pancreatic surgery for the first time in the world and further data was subsequently published by Horgan and his group at the University of Illinois and then later at the same institution by others. In 2007, the University of Illinois at Chicago medical team, led by Prof. Pier Cristoforo Giulianotti, reported a pancreatectomy and also the Midwest's first fully robotic Whipple surgery. In April 2008, the same team of surgeons performed the world's first fully minimally invasive liver resection for living donor transplantation, removing 60% of the patient's liver, yet allowing him to leave the hospital just a couple of days after the procedure, in very good condition. Furthermore, the patient can also leave with less pain than a usual surgery due to the four puncture holes and not a scar by a surgeon.

Cardiothoracic surgery

Robot-assisted MIDCAB and Endoscopic coronary artery bypass (TECAB) operations are being performed with the Da Vinci system. Mitral valve repairs and replacements have been performed. The Ohio State University, Columbus has performed CABG, mitral valve, esophagectomy, lung resection, tumor resections, among other robotic assisted procedures and serves as a training site for other surgeons. In 2002, surgeons at the Cleveland Clinic in Florida reported and published their preliminary experience with minimally invasive "hybrid" procedures. These procedures combined robotic revascularization and coronary stenting and further expanded the role of robots in coronary bypass to patients with disease in multiple vessels. Ongoing research on the outcomes of robotic assisted CABG and hybrid CABG is being done.

Cardiology and Electrophysiology

The Stereotaxis Magnetic Navigation System (MNS) has been developed to increase precision and safety in ablation procedures for arrhythmias and atrial fibrillation while reducing radiation exposure for the patient and physician, and the system utilizes two magnets to remotely steerable catheters. The system allows for automated 3-D mapping of the heart and vasculature, and MNS has also been used in interventional cardiology for guiding stents and leads in PCI and CTO procedures, proven to reduce contrast usage and access tortuous anatomy unreachable by manual navigation. Dr. Andrea Natale has referred to the new Stereotaxis procedures with the magnetic irrigated catheters as "revolutionary."

The Hansen Medical Sensei robotic catheter system uses a remotely operated system of pulleys to navigate a steerable sheath for catheter guidance. It allows precise and more forceful positioning of catheters used for 3-D mapping of the heart and vasculature. The system provides doctors with estimated force feedback information and feasible manipulation within the left atrium of the heart. The Sensei has been associated with mixed acute success rates compared to manual, commensurate with higher procedural complications, longer procedure times but lower fluoroscopy dosage to the patient.

At present, three types of heart surgery are being performed on a routine basis using robotic surgery systems. These three surgery types are:

- Atrial septal defect repair – the repair of a hole between the two upper chambers of the heart,

- Mitral valve repair – the repair of the valve that prevents blood from regurgitating back into the upper heart chambers during contractions of the heart,

- Coronary artery bypass – rerouting of blood supply by bypassing blocked arteries that provide blood to the heart.

As surgical experience and robotic technology develop, it is expected that the applications of robots in cardiovascular surgery will expand.

Colon and rectal surgery

Many studies have been undertaken in order to examine the role of robotic procedures in the field of colorectal surgery.

Results to date indicate that robotic-assisted colorectal procedures outcomes are "no worse" than the results in the now "traditional" laparoscopic colorectal operations. Robotic-assisted colorectal surgery appears to be safe as well. Most of the procedures have been performed for malignant colon and rectal lesions. However, surgeons are now moving into resections for diverticulitis and non-resective rectopexies (attaching the colon to the sacrum in order to treat rectal prolapse.)

When evaluated for several variables, robotic-assisted procedures fare equally well when compared with laparoscopic, or open abdominal operations. Study parameters have looked at intraoperative patient preparation time, length of time to perform the operation, adequacy of the removed surgical specimen with respect to clear surgical margins and number of lymph nodes removed, blood loss, operative or postoperative complications and long-term results.

More difficult to evaluate are issues related to the view of the operative field, the types of procedures that should be performed using robotic assistance and the potential added cost for a robotic operation.

Many surgeons feel that the optics of the 3-dimensional, two camera stereo optic robotic system are superior to the optical system used in laparoscopic procedures. The pelvic nerves are clearly visualized during robotic-assisted procedures. Less clear however is whether or not these supposedly improved optics and visualization improve patient outcomes with respect to postoperative impotence or incontinence, and whether long-term patient survival is improved by using the 3-dimensional optic system. Additionally, there is often a need for a wider, or "larger" view of the operative field than is routinely provided during robotic operations., The close-up view of the area under dissection may hamper visualization of the "bigger view", especially with respect to ureteral protection.

Questions remain unanswered, even after many years of experience with robotic-assisted colorectal operations. Ongoing studies may help clarify many of the issues of confusion associated with this novel surgical approach.

Gastrointestinal Surgery

Multiple types of procedures have been performed with either the 'Zeus' or da Vinci robot systems, including bariatric surgery and gastrectomy for cancer. Surgeons at various universities initially published case series demonstrating different techniques and the feasibility of GI surgery using the robotic devices. Specific procedures have been more fully evaluated, specifically esophageal fundoplication for the treatment of gastroesophageal reflux and Heller myotomy for the treatment of achalasia.

Other gastrointestinal procedures including colon resection, pancreatectomy, esophagectomy and robotic approaches to pelvic disease have also been reported.

Gynecology

Robotic surgery in gynecology is of uncertain benefit with it being unclear if it affects rates of complications. Gynecologic procedures may take longer with robot-assisted surgery but may be associated with a shorter hospital stay following hysterectomy. In the United States, robotic-assisted hysterectomy for benign conditions has been shown to be more expensive than conventional laparoscopic hysterectomy, with no difference in overall rates of complications.

This includes the use of the da Vinci surgical system in benign gynecology and gynecologic oncology. Robotic surgery can be used to treat fibroids, abnormal periods, endometriosis, ovarian tumors, uterine prolapse, and female cancers. Using the robotic system, gynecologists can perform hysterectomies, myomectomies, and lymph node biopsies.

Neurosurgery

Several systems for stereotactic intervention are currently on the market. The NeuroMate was the first neurosurgical robot, commercially available in 1997. Originally developed in Grenoble by Alim-Louis_Benabid's team, it is now owned by Renishaw. With installations in the United States, Europe and Japan, the system has been used in 8000 stereotactic brain surgeries by 2009. IMRIS Inc.'s SYMBIS(TM) Surgical System will be the version of NeuroArm, the world's first MRI-compatible surgical robot, developed for world-wide commercialization. Medtech's Rosa is being used by several institutions, including the Cleveland Clinic in the U.S, and in Canada at Sherbrooke University and the Montreal Neurological Institute and Hospital in Montreal (MNI/H). Between June 2011 and September 2012, over 150 neurosurgical procedures at the MNI/H have been completed robotized stereotaxy, including in the placement of depth electrodes in the treatment of epilepsy, selective resections, and stereotaxic biopsies.

Orthopedics

The ROBODOC system was released in 1992 by Integrated Surgical Systems, Inc. which merged into CUREXO Technology Corporation. Also, The Acrobot Company Ltd. developed the "Acrobot Sculptor", a robot that constrained a bone cutting tool to a pre-defined volume. The "Acrobot Sculptor" was sold to Stanmore Implants in August 2010. Stanmore received FDA clearance in February 2013 for US surgeries but sold the Sculptor to Mako Surgical in June 2013 to resolve a patent infringement lawsuit. Another example is the CASPAR robot produced by U.R.S.which is used for total hip replacement, total knee replacement and anterior cruciate ligament reconstruction. MAKO Surgical Corp (founded 2004) produces the RIO (Robotic Arm Interactive Orthopedic System) which combines robotics, navigation, and haptics for both partial knee and total hip replacement surgery. Blue Belt Technologies received

FDA clearance in November 2012 for the Navio™ Surgical System. The Navio System is a navigated, robotics-assisted surgical system that uses a CT free approach to assist in partial knee replacement surgery.

Pediatrics

Surgical robotics has been used in many types of pediatric surgical procedures including: tracheoesophageal fistula repair, cholecystectomy, nissen fundoplication, morgagni's hernia repair, kasai portoenterostomy, congenital diaphragmatic hernia repair, and others. On 17 January 2002, surgeons at Children's Hospital of Michigan in Detroit performed the nation's first advanced computer-assisted robot-enhanced surgical procedure at a children's hospital.

The Center for Robotic Surgery at Children's Hospital Boston provides a high level of expertise in pediatric robotic surgery. Specially-trained surgeons use a high-tech robot to perform complex and delicate operations through very small surgical openings. The results are less pain, faster recoveries, shorter hospital stays, smaller scars, and happier patients and families.

In 2001, Children's Hospital Boston was the first pediatric hospital to acquire a surgical robot. Today, surgeons use the technology for many procedures and perform more pediatric robotic operations than any other hospital in the world. Children's Hospital physicians have developed a number of new applications to expand the use of the robot, and train surgeons from around the world on its use.

Radiosurgery

The CyberKnife Robotic Radiosurgery System uses image guidance and computer controlled robotics to treat tumors throughout the body by delivering multiple beams of high-energy radiation to the tumor from virtually any direction. The system uses a German KUKA KR 240. Mounted on the robot is a compact X-band linac that produces 6MV X-ray radiation. Mounting the radiation source on the robot allows very fast repositioning of the source, which enables the system to deliver radiation from many different directions without the need to move both the patient and source as required by current gantry configurations.

Transplant Surgery

Transplant surgery (organ transplantation) has been considered as highly technically demanding and virtually unobtainable by means of conventional laparoscopy. For many years, transplant patients were unable to benefit from the advantages of minimally invasive surgery. The development of robotic technology and its associated high resolution capabilities, three dimensional visual system, wrist type motion and fine instruments, gave opportunity for highly complex procedures to be completed in a minimally invasive fashion. Subsequently, the first fully robotic kidney transplantations were performed in the late 2000s. After the procedure was proven to be feasible and safe, the main emerging challenge was to determine which patients would benefit most from this robotic technique. As a result, recognition of the increasing prevalence of obesity amongst patients with kidney failure on hemodialysis posed a significant problem. Due to the abundantly higher risk of complications after traditional open kidney transplantation, obese patients were frequently denied access to transplantation, which is the premium treatment for end stage kidney disease. The use of the robotic-assisted approach has allowed kidneys to be transplanted with minimal incisions, which has virtually alleviated wound complications and significantly shortened the recovery period. The University of Illinois Medical Center reported the largest series of 104 robotic-assisted kidney transplants for obese recipients (mean body mass index > 42). Amongst this group of patients, no wound infections were observed and the function of transplanted kidneys was excellent. In this way, robotic kidney transplantation could be considered as the biggest advance in surgical technique for this procedure since its creation more than half a century ago.

Urology

Robotic surgery in the field of urology has become very popular, especially in the United States. It has been most extensively applied for excision of prostate cancer because of difficult anatomical access. It is also utilized for kidney cancer surgeries and to lesser extent surgeries of the bladder.

As of 2014, there is little evidence of increased benefits compared to standard surgery to justify the increased costs. Some have found tentative evidence of more complete removal of cancer and less side effects from surgery for prostatectomy.

In 2000, the first robot-assisted laparoscopic radical prostatectomy was performed.

Vascular Surgery

In September 2010, the first robotic operations with Hansen Medical's Magellan Robotic System at the femoral vasculature were performed at the University Medical Centre Ljubljana (UMC Ljubljana), Slovenia. The research was led by Borut Geršak, the head of the Department of Cardiovascular Surgery at the centre. Geršak explained that the robot used was the first true robot in the history of robotic surgery, meaning the user interface was not resembling surgical instruments and the robot was not simply imitating the movement of human hands but was guided by pressing buttons, just like one would play a video game. The robot was imported to Slovenia from the United States.

Miniature Robotics

As scientists seek to improve the versatility and utility of robotics in surgery, some are attempting to miniaturize the robots. For example, the University of Nebraska Medical Center has led a multi-campus effort to provide collaborative research on mini-robotics among surgeons, engineers and computer scientists.

History

The first robot to assist in surgery was the *Arthrobot*, which was developed and used for the first time in Vancouver in 1983. Intimately involved were biomedical engineer, Dr. James McEwen, Geof Auchinleck, a UBC engineering physics grad, and Dr. Brian Day as well as a team of engineering students. The robot was used in an orthopaedic surgical procedure on 12 March 1984, at the UBC Hospital in Vancouver. Over 60 arthroscopic surgical procedures were performed in the first 12 months, and a 1985 National Geographic video on industrial robots, *The Robotics Revolution*, featured the device. Other related robotic devices developed at the same time included a surgical scrub nurse robot, which handed operative instruments on voice command, and a medical laboratory robotic arm. A YouTube video entitled *Arthrobot* illustrates some of these in operation.

In 1985 a robot, the Unimation Puma 200, was used to place a needle for a brain biopsy using CT guidance. In 1992, the PROBOT, developed at Imperial College London, was used to perform prostatic surgery by Dr. Senthil Nathan at Guy's and St Thomas' Hospital, London. This was the first pure robotic surgery in the world. Also the Robot Puma 560, a robot developed in 1985 by Kwoh et al. Puma 560 was used to perform neurosurgical biopsies with greater precision. Just like with any other technological innovation, this system led to the development of new and improved surgical robot called PROBOT. The PROBOT was specifically designed for transurethral resection of the prostate. Meanwhile, when PROBOT was being developed, ROBODOC, a robotic system designed to assist hip replacement surgeries was the first surgical robot that was approved by the FDA. The ROBODOC from Integrated Surgical Systems (working closely with IBM) was introduced in 1992 to mill out precise fittings in the femur for hip replacement. The purpose of the ROBODOC was to replace the previous method of carving out a femur for an implant, the use of a mallet and broach/rasp.

Further development of robotic systems was carried out by SRI International and Intuitive Surgical with the introduction of the da Vinci Surgical System and Computer Motion with the *AESOP* and the ZEUS robotic surgical system. The first robotic surgery took place at The Ohio State University Medical Center in Columbus, Ohio under the direction of Robert E. Michler. Examples of using ZEUS include a fallopian tube reconnection in July 1998, a *beating heart* coronary artery bypass graft in October 1999, and the Lindbergh Operation, which was a cholecystectomy performed remotely in September 2001.

The original telesurgery robotic system that the da Vinci was based on was developed at SRI International in Menlo Park with grant support from DARPA and NASA. Although the telesurgical robot was originally intended to facilitate remotely performed surgery in battlefield and other remote environments, it turned out to be more useful for minimally invasive on-site surgery. The patents for the early prototype were sold to Intuitive Surgical in Mountain View, California. The da Vinci senses the surgeon's hand movements and translates them electronically into scaled-down micro-movements to manipulate the tiny proprietary instruments. It also detects and filters out any tremors in the surgeon's hand movements, so that they are not duplicated robotically. The camera used in the system provides a true stereoscopic picture transmitted to a surgeon's console. Examples of using the da Vinci system include the first robotically assisted heart bypass (performed in Germany) in May 1998, and the first performed in the United States in September 1999; and the first all-robotic-assisted kidney transplant, performed in January 2009. The da Vinci Si was released in April 2009, and initially sold for $1.75 million.

In May 2006 the first artificial intelligence doctor-conducted unassisted robotic surgery on a 34-year-old male to correct heart arythmia. The results were rated as better than an above-average human surgeon. The machine had a database of 10,000 similar operations, and so, in the words of its designers, was "more than qualified to operate on any patient". In August 2007, Dr. Sijo Parekattil of the Robotics Institute and Center for Urology (Winter Haven Hospital and University of Florida) performed the first robotic assisted microsurgery procedure denervation of the spermatic cord for chronic testicular pain. In February 2008, Dr. Mohan S. Gundeti of the University of Chicago Comer Children's Hospital performed the first robotic pediatric neurogenic bladder reconstruction.

On 12 May 2008, the first image-guided MR-compatible robotic neurosurgical procedure was performed at University of Calgary by Dr. Garnette Sutherland using the NeuroArm. In June 2008, the German Aerospace Centre (DLR) presented a robotic system for minimally invasive surgery, the MiroSurge. In September 2010, the Eindhoven University of Technology announced the development of the Sofie surgical system, the first surgical robot to employ force feedback. In September 2010, the first robotic operation at the femoral vasculature was performed at the University Medical Centre Ljubljana by a team led by Borut Geršak.

Parallel Manipulator

A parallel manipulator is a mechanical system that uses several computer-controlled serial chains to support a single platform, or end-effector. Perhaps, the best known parallel manipulator is formed from six linear actuators that support a movable base for devices such as flight simulators. This device is called a Stewart platform or the Gough-Stewart platform in recognition of the engineers who first designed and used them.

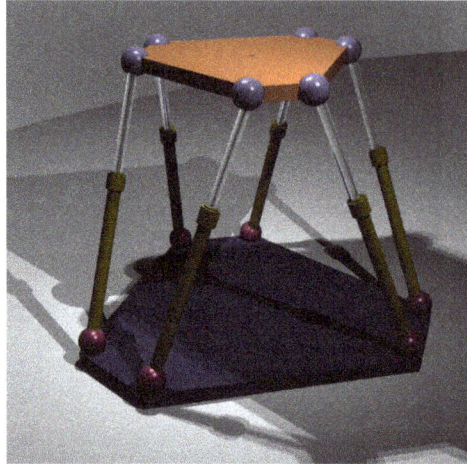

Abstract render of a Hexapod platform (Stewart Platform)

Also known as parallel robots, or generalized Stewart platforms (in the Stewart platform, the actuators are paired together on both the basis and the platform), these systems are articulated robots that use similar mechanisms for the movement of either the robot on its base, or one or more manipulator arms. Their 'parallel' distinction, as opposed to a serial manipulator, is that the end effector (or 'hand') of this linkage (or 'arm') is connected to its base by a number of (usually three or six) separate and independent linkages working in parallel. 'Parallel' is used here in the computer science sense, rather than the geometrical; these linkages act together, but it is not implied that they are aligned as parallel lines; here *parallel* means that the position of the end point of each linkage is independent of the position of the other linkages.

Design Features

A parallel manipulator is designed so that each chain is usually short, simple and can thus be rigid against unwanted movement, compared to a serial manipulator. Errors in one chain's positioning are averaged in conjunction with the others, rather than being cumulative. Each actuator must still move within its own degree of freedom, as for a serial robot; however in the parallel robot the off-axis flexibility of a joint is also constrained by the effect of the other chains. It is this closed-loop stiffness that makes the overall parallel manipulator stiff relative to its components, unlike the serial chain that becomes progressively less rigid with more components.

This mutual stiffening also permits simple construction: Stewart platform hexapods chains use prismatic joint linear actuators between any-axis universal ball joints. The ball joints are passive: simply free to move, without actuators or brakes; their position is constrained solely by the other chains. Delta robots have base-mounted rotary actuators that move a light, stiff, parallelogram arm. The effector is mounted between the tips of three of these arms and again, it may be mounted with simple ball-joints. Static representation of a parallel robot is often akin to that of a pin-jointed truss: the links and their actuators feel only tension or compression, without any bending or torque, which again reduces the effects of any flexibility to off-axis forces.

A further advantage of the parallel manipulator is that the heavy actuators may often be centrally mounted on a single base platform, the movement of the arm taking place through struts and joints alone. This reduction in mass along the arm permits a lighter arm construction, thus lighter actuators and faster movements. This centralisation of mass also reduces the robot's overall moment of inertia, which may be an advantage for a mobile or walking robot.

All these features result in manipulators with a wide range of motion capability. As their speed of action is often constrained by their rigidity rather than sheer power, they can be fast-acting, in comparison to serial manipulators.

Comparison to Serial Manipulators

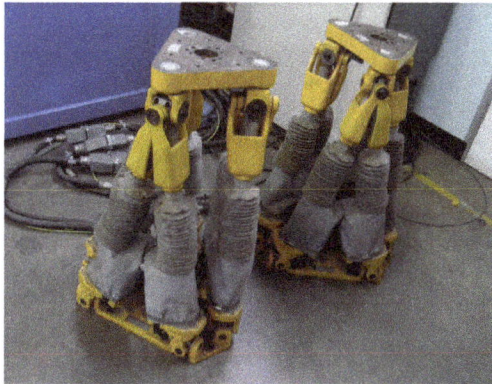

Hexapod positioning systems, also known as Stewart Platforms.

Most robot applications require rigidity. Serial robots may achieve this by using high-quality rotary joints that permit movement in one axis but are rigid against movement outside this. Any joint permitting movement *must* also have this movement under deliberate control by an actuator. A movement requiring several axes thus requires a number of such joints. Unwanted flexibility or sloppiness in one joint causes a similar sloppiness in the arm, which may be amplified by the distance between the joint and the end-effectuor: there is no opportunity to brace one joint's movement against another. Their inevitable hysteresis and off-axis flexibility accumulates along the arm's kinematic chain; a precision serial manipulator is a compromise between precision, complexity, mass (of the manipulator and of the manipulated objects) and cost. On the other hand, with parallel manipulators, a high rigidity may be obtained with a small mass of the manipulator (relatively to the charge being manipulated). This allows high precision and high speed of movements, and motivates the use of parallel manipulators in flight simulators (high speed with rather large masses) and electrostatic or magnetic lenses in particle accelerators (very high precision in positioning large masses).

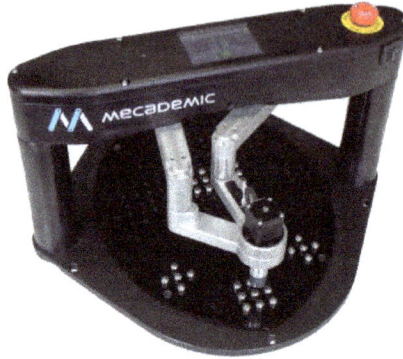

A five-bar parallel robot

A drawback of parallel manipulators, in comparison to serial manipulators, is their limited workspace. As for serial manipulators, the workspace is limited by the geometrical and mechanical limits of the design (collisions between legs maximal and minimal lengths of the legs). The workspace is also limited by the existence of *singularities*, which are positions where, for some trajectories of the movement, the variation of the lengths of the legs is infinitely smaller than the variation of the position. Conversely, at a singular position, a force (like gravity) applied on the end-effector induce infinitely large constraints on the legs, which may result in a kind of "explosion" of the manipulator. The determination of the singular positions is difficult (for a general parallel manipulator, this is an open problem). This implies that the workspaces of the parallel manipulators are, usually, artificially limited to a small region where one knows that there is no singularity.

Another drawback of parallel manipulators is their nonlinear behavior: the command which is needed for getting a linear or a circular movement of the end-effector depends dramatically on the location in the workspace and does not vary linearly during the movement. Because of the difficulty of such a non-linear command, the parallel manipulators are not yet used in high precision machining, despite their excellent mechanical properties (speed and precision).

Applications

Major industrial applications of these devices are:

- flight simulators

- automobile simulators

- in work processes

- photonics / optical fiber alignment

They also become more popular:

- in high speed, high-accuracy positioning with limited workspace, such as in assembly of PCBs

- as micro manipulators mounted on the end effector of larger but slower serial manipulators

- as high speed/high-precision milling machines

Parallel robots are usually more limited in the workspace; for instance, they generally cannot reach around obstacles. The calculations involved in performing a desired manipulation (forward kinematics) are also usually more difficult and can lead to multiple solutions.

Two examples of popular parallel robots are the Stewart platform and the Delta robot.

Delta Robot

Sketchy, a portrait-drawing delta robot

A delta robot is a type of parallel robot. It consists of three arms connected to universal joints at the base. The key design feature is the use of parallelograms in the arms, which maintains the orientation of the end effector. By contrast, a Stewart platform can change the orientation of its end effector.

Delta robots have popular usage in picking and packaging in factories because they can be quite fast, some executing up to 300 picks per minute.

History

The delta robot (a parallel arm robot) was invented in the early 1980s by a research team led by professor Reymond Clavel at the École Polytechnique Fédérale de Lausanne

(EPFL, Switzerland). The purpose of this new type of robot was to manipulate light and small objects at a very high speed, an industrial need at that time. In 1987, the Swiss company Demaurex purchased a license for the delta robot and started the production of delta robots for the packaging industry. In 1991 Reymond Clavel presented his doctoral thesis 'Conception d'un robot parallèle rapide à 4 degrés de liberté', and received the golden robot award in 1999 for his work and development of the delta robot. Also in 1999, ABB Flexible Automation started selling its delta robot, the FlexPicker. By the end of 1999 delta robots were also sold by Sigpack Systems.

Commercial pick and place robots

In 2009, FANUC released the newest version of the Delta robot, the FANUC M-1iA Robot, and would later release variations of this Delta robot for heavier payloads. FANUC released the M-3iA in 2010 for heavier payloads, and most recently the FANUC M-2iA Robot for medium-sized payloads in 2012.

Design

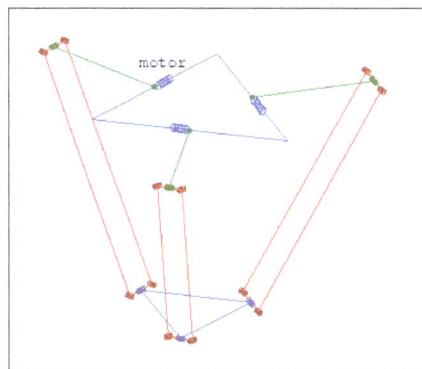

Delta robot kinematics

The delta robot is a parallel robot, i.e. it consists of multiple kinematic chains connecting the base with the end-effector. The robot can also be seen as a spatial generalisation of a four-bar linkage.

The key concept of the delta robot is the use of parallelograms which restrict the movement of the end platform to pure translation, i.e. only movement in the X, Y or Z direction with no rotation.

The robot's base is mounted above the workspace and all the actuators are located on it. From the base, three middle jointed arms extend. The ends of these arms are connected to a small triangular platform. Actuation of the input links will move the triangular platform along the X, Y or Z direction. Actuation can be done with linear or rotational actuators, with or without reductions (direct drive).

Since the actuators are all located in the base, the arms can be made of a light composite material. As a result of this, the moving parts of the delta robot have a small inertia. This allows for very high speed and high accelerations. Having all the arms connected together to the end-effector increases the robot stiffness, but reduces its working volume.

The version developed by Reymond Clavel has four degrees of freedom: three translations and one rotation. In this case a fourth leg extends from the base to the middle of the triangular platform giving to the end effector a fourth, rotational degree of freedom around the vertical axis.

Currently other versions of the delta robot have been developed:

- Delta with 6 degrees of freedom: developed by the company Fanuc, on which a serial kinematic with 3 rotational degrees of freedom is placed on the end effector

- Delta with 4 degrees of freedom: developed by the company Adept, which has 4 parallelogram directly connected to the end-platform instead of having a fourth leg coming in the middle of the end-effector

- Pocket Delta: developed by the Swiss company Astral SA, a 3-axis version of the Delta Robot adapted for flexible part feeding systems and other high-speed, high-precision applications.

- Delta Direct Drive: a 3 degrees of freedom Delta Robot having the motor directly connected to the arms. Accelerations can be very high, from 30 up to 100 g.

- Delta Cube: developed by the EPFL university laboratory LSRO, a delta robot built in a monolithic design, having flexure-hinges joints. This robot is adapted for ultra-high-precision applications.

- Several "linear delta" arrangements have been developed where the motors drive linear actuators rather than rotating an arm. Such linear delta arrangements can have much larger working volumes than rotational delta arrangements.

The majority of delta robots use rotary actuators. Vertical linear actuators have recently been used (using a linear delta design) to produce a novel design of 3D printer. These offer advantages over conventional leadscrew-based 3D printers of quicker access to a larger build volume for a comparable investment in hardware.

Applications

Large delta-style 3D printer

Industries that take advantage of the high speed of delta robots are the packaging industry, medical and pharmaceutical industry. For its stiffness it is also used for surgery. Other applications include high precision assembly operations in a clean room for electronic components. The structure of a delta robot can also be used to create haptic controllers. More recently, the technology has been adapted to 3D printers. These printers can be built for about a thousand dollars and compete well with the traditional Cartesian printers from the RepRap project.

Pneumatic Cylinder

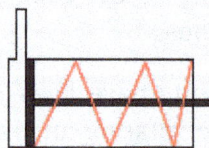

Operation diagram of a single acting cylinder. The spring (red) can also be outside the cylinder, attached to the item being moved.

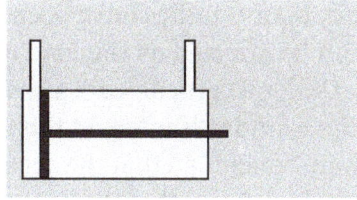

Operation diagram of a double acting cylinder

3D animated pneumatic cylinder (CAD)

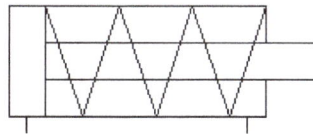

Schematic symbol for pneumatic cylinder with spring return

Pneumatic cylinder(s) (sometimes known as air cylinders) are mechanical devices which use the power of compressed gas to produce a force in a reciprocating linear motion.

Like hydraulic cylinders, something forces a piston to move in the desired direction. The piston is a disc or cylinder, and the piston rod transfers the force it develops to the object to be moved. Engineers sometimes prefer to use pneumatics because they are quieter, cleaner, and do not require large amounts of space for fluid storage.

Because the operating fluid is a gas, leakage from a pneumatic cylinder will not drip out and contaminate the surroundings, making pneumatics more desirable where cleanliness is a requirement. For example, in the mechanical puppets of the Disney Tiki Room, pneumatics are used to prevent fluid from dripping onto people below the puppets.

Operation

General

Once actuated, compressed air enters into the tube at one end of the piston and, hence, imparts force on the piston. Consequently, the piston becomes displaced.

Compressibility of Gasses

One major issue engineers come across working with pneumatic cylinders has to do with the compressibility of a gas. Many studies have been completed on how the precision of a pneumatic cylinder can be affected as the load acting on the cylinder tries to further compress the gas used. Under a vertical load, a case where the cylinder takes on the full load, the precision of the cylinder is affected the most. A study at the National Cheng Kung University in Taiwan, concluded that the accuracy is about ± 30 nm, which is still within a satisfactory range but shows that the compressibility of air has an effect on the system.

Fail Safe Mechanisms

Pneumatic systems are often found in settings where even rare and brief system failure is unacceptable. In such situations locks can sometimes serve as a safety mechanism in case of loss of air supply (or its pressure falling) and, thus remedy or abate any damage arising in such a situation. Leakage of air from the input or output reduces the pressure and so the desired output.

Types

Although pneumatic cylinders will vary in appearance, size and function, they generally fall into one of the specific categories shown below. However, there are also numerous other types of pneumatic cylinder available, many of which are designed to fulfill specific and specialized functions.

Single-acting Cylinders

Single-acting cylinders (SAC) use the pressure imparted by compressed air to create a driving force in one direction (usually out), and a spring to return to the "home" position. More often than not, this type of cylinder has limited extension due to the space the compressed spring takes up. Another downside to SACs is that part of the force produced by the cylinder is lost as it tries to push against the spring.

Double-acting Cylinders

Double-acting cylinders (DAC) use the force of air to move in both extend and retract strokes. They have two ports to allow air in, one for outstroke and one for instroke. Stroke length for this design is not limited, however, the piston rod is more vulnerable to buckling and bending. Additional calculations should be performed as well.

Multi-stage, Telescoping Cylinder

Telescoping cylinders, also known as telescopic cylinders can be either single or double-acting. The telescoping cylinder incorporates a piston rod nested within a series of hollow stages of increasing diameter. Upon actuation, the piston rod and each succeeding stage "telescopes" out as a segmented piston. The main benefit of this design is the allowance for a notably longer stroke than would be achieved with a single-stage cylinder of the same collapsed (retracted) length. One cited drawback to telescoping cylinders is the increased potential for piston flexion due to the segmented piston design. Consequently, telescoping cylinders are primarily utilized in applications where the piston bears minimal side loading.

pneumatic telescoping cylinder, 8-stages, single-acting, retracted and extended

Other Types

Although SACs and DACs are the most common types of pneumatic cylinder, the following types are not particularly rare:

- Through rod air cylinders: piston rod extends through both sides of the cylinder, allowing for equal forces and speeds on either side.

- Cushion end air cylinders: cylinders with regulated air exhaust to avoid impacts between the piston rod and the cylinder end cover.

- Rotary air cylinders: actuators that use air to impart a rotary motion.

- Rodless air cylinders: These have no piston rod. They are actuators that use a mechanical or magnetic coupling to impart force, typically to a table or other body that moves along the length of the cylinder body, but does not extend beyond it.

- Tandem air cylinder: two cylinders are assembled in series in order to double the force output.

- Impact air cylinder: high velocity cylinders with specially designed end covers that withstand the impact of extending or retracting piston rods.

Rodless Cylinders

Some rodless types have a slot in the wall of the cylinder that is closed off for much of its length by two flexible metal sealing bands. The inner one prevents air from escaping, while the outer one protects the slot and inner band. The piston is actually a pair of them, part of a comparatively long assembly. They seal to the bore and inner band at both ends of the assembly. Between the individual pistons, however, are camming surfaces that "peel off" the bands as the whole sliding assembly moves toward the sealed volume, and "replace" them as the assembly moves away from the other end. Between the camming surfaces is part of the moving assembly that protrudes through the slot to move the load. Of course, this means that the region where the sealing bands are not in contact is at atmospheric pressure.

Another type has cables (or a single cable) extending from both (or one) end[s] of the cylinder. The cables are jacketed in plastic (nylon, in those referred to), which provides a smooth surface that permits sealing the cables where they pass through the ends of the cylinder. Of course, a single cable has to be kept in tension.

Still others have magnets inside the cylinder, part of the piston assembly, that pull along magnets outside the cylinder wall. The latter are carried by the actuator that moves the load. The cylinder wall is thin, to ensure that the inner and outer magnets are near each other. Multiple modern high-flux magnet groups transmit force without disengaging or excessive resilience.

Design

Construction

Depending on the job specification, there are multiple forms of body constructions available:

- Tie rod cylinders: The most common cylinder constructions that can be used in many types of loads. Has been proven to be the safest form.

- Flanged-type cylinders: Fixed flanges are added to the ends of cylinder, however, this form of construction is more common in hydraulic cylinder construction.

- One-piece welded cylinders: Ends are welded or crimped to the tube, this form is inexpensive but makes the cylinder non-serviceable.

- Threaded end cylinders: Ends are screwed onto the tube body. The reduction of material can weaken the tube and may introduce thread concentricity problems to the system.

Material

Upon job specification, the material may be chosen. Material range from nickel-plated brass to aluminum, and even steel and stainless steel. Depending on the level of loads, humidity, temperature, and stroke lengths specified, the appropriate material may be selected.

Mounts

Depending on the location of the application and machinability, there exist different kinds of mounts for attaching pneumatic cylinders:

Type of Mount Ends	
Rod End	**Cylinder End**
Plain	Plain
Threaded	Foot
Clevis	Bracket-single or double
Torque or eye	Trunnion
Flanged	Flanged
	Clevis etc.

Sizes

Air cylinders are available in a variety of sizes and can typically range from a small 2.5 mm ($\frac{1}{10}$ in) air cylinder, which might be used for picking up a small transistor or other electronic component, to 400 mm (16 in) diameter air cylinders which would impart enough force to lift a car. Some pneumatic cylinders reach 1,000 mm (39 in) in diameter, and are used in place of hydraulic cylinders for special circumstances where leaking hydraulic oil could impose an extreme hazard.

Pressure, Radius, Area and Force Relationships

Rod Stresses

Due to the forces acting on the cylinder, the piston rod is the most stressed component and has to be designed to withstand high amounts of bending, tensile and compressive forces. Depending on how long the piston rod is, stresses can be calculated differently. If the rods length is less than 10 times the diameter, then it may be treated as a rigid body which has compressive or tensile forces acting on it. In which case the relationship is:

$$F = A\sigma$$

Where:

F is the compressive or tensile force

A is the cross-sectional area of the piston rod

σ is the stress

However, if the length of the rod exceeds the 10 times the value of the diameter, then the rod needs to be treated as a column and buckling needs to be calculated as well.

Instroke and Outstroke

Although the diameter of the piston and the force exerted by a cylinder are related, they are not directly proportional to one another. Additionally, the typical mathematical relationship between the two assumes that the air supply does not become saturated. Due to the effective cross sectional area reduced by the area of the piston rod, the instroke force is less than the outstroke force when both are powered pneumatically and by same supply of compressed gas.

The relationship between the force, radius, and pressure can derived from simple distributed load equation:

$$F_r = PA_e$$

Where:

F_r is the resultant force

P is the pressure or distributed load on the surface

A_e is the effective cross sectional area the load is acting on

Outstroke

Using the distributed load equation provided the A_e can be replaced with area of the piston surface where the pressure is acting on.

$$F_r = P(\pi r^2)$$

Where:

F_r represents the resultant force

r represents the radius of the piston

π is pi, approximately equal to 3.14159.

Instroke

On instroke, the same relationship between force exerted, pressure and *effective cross sectional area* applies as discussed above for outstroke. However, since the cross sectional area is less than the piston area the relationship between force, pressure and *radius* is different. The calculation isn't more complicated though, since the effective cross sectional area is merely that of the piston surface minus the cross sectional area of the piston rod.

For instroke, therefore, the relationship between force exerted, pressure, radius of the piston, and radius of the piston rod, is as follows:

$$F_r = P(\pi r_1^2 - \pi r_2^2) = P\pi(r_1^2 - r_2^2)$$

Where:

F_r represents the resultant force

r_1 represents the radius of the piston

r_2 represents the radius of the piston rod

π is pi, approximately equal to 3.14159.

Mobile Servicing System

Astronaut Stephen K. Robinson anchored to the end of *Canadarm2* during STS-114

The Mobile Servicing System (MSS), is a robotic system on board the International Space Station (ISS). Launched to the ISS in 2001, it plays a key role in station assembly and maintenance; it moves equipment and supplies around the station, supports astronauts working in space, and services instruments and other payloads attached to the ISS and is used for external maintenance. Astronauts receive specialized training to enable them to perform these functions with the various systems of the MSS.

The MSS is composed of three components - the Space Station Remote Manipulator System (SSRMS), known as Canadarm2, the Mobile Remote Servicer Base System (MBS) and the Special Purpose Dexterous Manipulator (SPDM, also known as *Dextre* or *Canada hand*). The system can move along rails on the Integrated Truss Structure on top of the US provided Mobile Transporter cart which hosts the MRS Base System. The system's control software was written in the Ada 95 programming language.

Canadarm2 moves *Rassvet* to berth with the station on STS-132.

The MSS was designed and manufactured by MDA Space Missions (previously called MD Robotics; previously called SPAR Aerospace) for the Canadian Space Agency's contribution to the International Space Station.

Canadarm2

Astronaut Leroy Chiao works with the control of the *Canadarm2* in the Destiny lab

The Canadian Government logo featured on the side of the *Canadarm2*.

Launched on STS-100 in April 2001, this second generation arm is a larger, more advanced version of the Space Shuttle's original Canadarm. *Canadarm2* is 17.6 m (58 ft) when fully extended and has seven motorized joints. It has a mass of 1,800 kg (4,000 lb) and a diameter of 35 cm (14 in). The arm is capable of handling large payloads of up to 116,000 kg (256,000 lb) and was able to assist with docking the space shuttle. Officially known as the *Space Station Remote Manipulator System* (SSRMS), it is self-relocatable and can move end-over-end to reach many parts of the Space Station in an inchworm-like movement. In this movement, it is limited only by the number of Power Data Grapple Fixtures (PDGFs) on the station. PDGFs located around the station provide power, data and video to the arm through its Latching End Effectors (LEEs). The arm can also travel the entire length of the space station truss using the Mobile Base System.

In addition to moving itself around the station, the arm can move any object with a grapple fixture. In construction of the station the arm was used to move large segments into place. It can also be used to capture unpiloted ships like the SpaceX Dragon, Cygnus (spacecraft) and Japanese H-II Transfer Vehicle(HTV) which are equipped with a standard grapple fixture which the Canadarm2 uses to capture and dock the spacecraft. The arm is also used to undock and release the spacecraft after use.

On-board operators see what they are doing by looking at the three Robotic Work Station (RWS) LCD screens. The MSS has two RWS units: one located in the Destiny module (US Lab module) and the other in the Cupola. Only one RWS controls the MSS at a time. The RWS has two sets of control joysticks: one Rotational Hand Controller (RHC) and one Translational Hand Controller (THC). In addition to this is the Display and Control Panel (DCP) and the Portable Computer System (PCS) laptop.

In recent years, the majority of robotic operations are commanded remotely by flight controllers on the ground at Mission Control Center (NASA), or from the Canadian Space Agency . Operators can work in shifts to accomplish objectives with more flexibility than when done by on-board crew operators, albeit at a slower pace. Astronaut operators are used for time-critical operations such as visiting vehicle captures and robotics supported Extra-Vehicular Activity.

Special Purpose Dexterous Manipulator

Dextre and Canadarm2 docked side by side on Power Data Grapple Fixtures

The Special Purpose Dexterous Manipulator, or "Dextre", is a smaller two-armed robot that can attach to Canadarm2, the ISS or the Mobile Base System. The arms and its power tools are capable of handling the delicate assembly tasks and changing Orbital Replacement Units (ORUs) currently handled by astronauts during space walks. Although Canadarm2 can move around the station in an "inchworm motion", it's unable to carry anything with it unless Dextre is attached. Testing was done in the space simulation chambers of the Canadian Space Agency's David Florida Laboratory in Ottawa. The manipulator was launched to the station 11 March 2008 on STS-123.

Mobile Base System

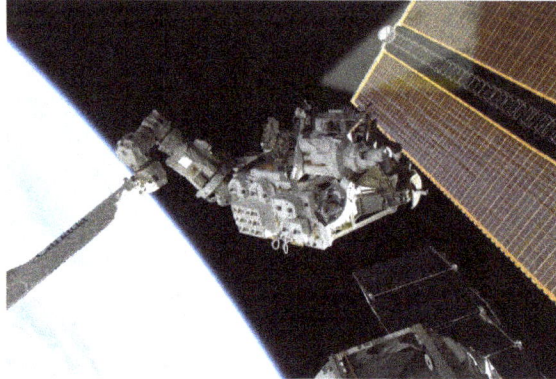

The Mobile Base System just before *Canadarm2* installed it on the Mobile Transporter during STS-111

The Mobile Remote Servicer Base System (MBS) is a base platform for the robotic arms. It was added to the station during STS-111 in June 2002. The platform rests atop the Mobile Transporter (installed on STS-110, designed by Northrop Grumman in Carpinteria, CA), which allows it to glide 108 metres down rails on the station's main truss. Canadarm2 can relocate by itself, but can't carry at the same time, Dextre can't relocate by itself. The MBS gives the two robotic arms the ability to travel to work sites all along the truss structure and to step off onto grapple fixtures along the way. When *Canadarm2* and *Dextre* are attached to the MBS, they have a combined mass of 4,900 kg (10,800 lb). Like *Canadarm2* it was built by MD Robotics and it has a minimum service life of 15 years.

The MBS is equipped with 4 Power Data Grapple Fixtures, one at each of its four top corners. Any of these can be used as a base for the two robots, *Canadarm2* and *Dextre*, as well as any of the payloads that might be held by them. The MBS also has 2 locations to attach payloads. The first is the *Payload/Orbital Replacement Unit Accommodations* (POA). This is a device that looks and functions much like the Latching End Effectors of *Canadarm2*. It can be used to park, power and command any payload with a grapple fixture, while keeping *Canadarm2* free to do something else. The other attachment location is the MBS Common Attachment System (MCAS). This is another type of attachment system that is used to host scientific experiments.

The MBS also supports astronauts during Extra-vehicular activities. It has locations to store tools and equipment, foot-restraints, handrails and safety tether attachment points as well as a camera assembly. If needed, it is even possible for an astronaut to "ride" the MBS while it moves at a top speed of about 1.5 meters per minute.

Canadarm2 riding the Mobile Base System along the Mobile Transporter railway,
running the length of the station's main truss

Enhanced ISS Boom Assembly

Installed on May 27, 2011, is a 50-foot boom with handrails and inspection cameras,
attached to the end of Canadarm2.

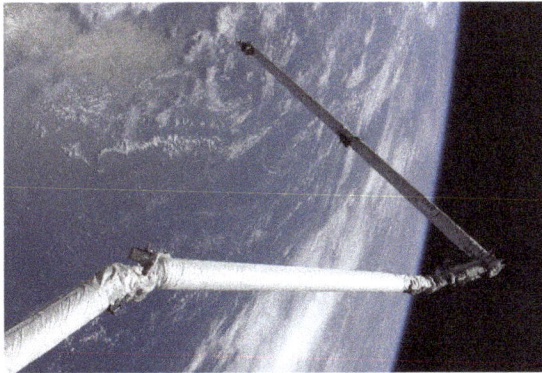

Shuttle Remote Manipulator System (RMS) holding OBSS boom on STS-114

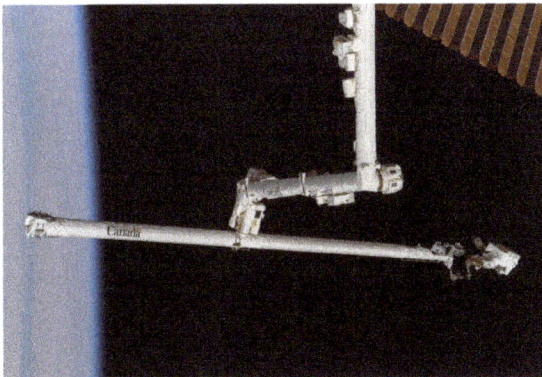

Astronaut Scott Parazynski (at right) riding the OBSS boom to repair the solar array during STS-120

Other ISS robotics

The station received a second robotic arm during STS-124, the Japanese Experiment
Module Remote Manipulator System (JEM-RMS). The JEM-RMS will be primarily
used to service the JEM Exposed Facility. An additional robotic arm, the European

Robotic Arm (ERA) is scheduled to launch alongside the Russian-built Multipurpose Laboratory Module during 2017.

Connected to Pirs, the ISS also has two Strela cargo cranes. One of the cranes can be extended to reach the end of Zarya. The other can extend to the opposite site and reach the end of Zvezda. The first crane was assembled in space during STS-96 and STS-101. The second crane was launched alongside Pirs itself.

Cartesian Coordinate Robot

A cartesian coordinate robot (also called linear robot) is an industrial robot whose three principal axis of control are linear (i.e. they move in a straight line rather than rotate) and are at right angles to each other.The three sliding joints correspond to moving the wrist up-down,in-out,back-forth. Among other advantages, this mechanical arrangement simplifies the Robot control arm solution. Cartesian coordinate robots with the horizontal member supported at both ends are sometimes called Gantry robots; mechanically, they resemble gantry cranes, although the latter are not generally robots. Gantry robots are often quite large.

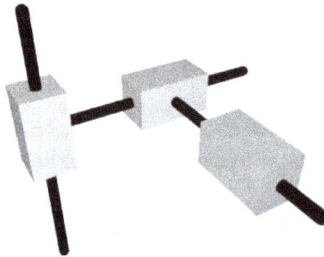

Kinematic diagram of cartesian coordinate robot

a plotter is an implemention of the cartesian coordinate robot

A popular application for this type of robot is a computer numerical control machine

(CNC machine) and 3D printing. The simplest application is used in milling and drawing machines where a pen or router translates across an x-y plane while a tool is raised and lowered onto a surface to create a precise design. Pick and place machines and plotters are also based on the principal of the cartesian coordinate robot.

References

- Needham, Joseph (1991). Science and Civilisation in China: Volume 2, History of Scientific Thought. Cambridge University Press. ISBN 0-521-05800-7.

- Rosheim, Mark E. (1994). Robot Evolution: The Development of Anthrobotics. Wiley-IEEE. pp. 9–10. ISBN 0-471-02622-0.

- Crane, Carl D.; Joseph Duffy (1998). Kinematic Analysis of Robot Manipulators. Cambridge University Press. ISBN 0-521-57063-8. Retrieved 2007-10-16.

- "Focal Points Seminar on review articles in the future of work - Safety and health at work - EU-OSHA". osha.europa.eu. Retrieved 2016-04-19.

- "iSplash-II: Realizing Fast Carangiform Swimming to Outperform a Real Fish" (PDF). Robotics Group at Essex University. Retrieved 2015-09-29.

- "iSplash-I: High Performance Swimming Motion of a Carangiform Robotic Fish with Full-Body Coordination" (PDF). Robotics Group at Essex University. Retrieved 2015-09-29.

- "Hoosier Daddy – The Largest Delta 3D Printer In the World". 3D Printer World. Punchbowl Media. 23 July 2014. Retrieved 28 September 2014.

- "Robotics Degree Programs at Worcester Polytechnic Institute". Worcester Polytechnic Institute. 2013. Retrieved 2013-04-12.

- Toy, Tommy (June 29, 2011). "Outlook for robotics and Automation for 2011 and beyond are excellent says expert". PBT Consulting. Retrieved 2012-01-27.

- "V UKC Ljubljana prvič na svetu uporabili žilnega robota za posege na femoralnem žilju" [The First Use of a Vascular Robot for Procedures on Femoral Vasculature] (in Slovenian). 8 November 2010. Retrieved 1 April 2011.

- "Robotics: the Future of Minimally Invasive Heart Surgery". Biomed.brown.edu. 6 October 1999. Retrieved 29 November 2011.

- "Surgeons perform world's first pediatric robotic bladder reconstruction". Esciencenews.com. 20 November 2008. Retrieved 29 November 2011.

Innovations of Mechatronics and Robotics

Mechatronics and robotics together have made possible such ground-breaking inventions like Mars Exploration Rover, Curiosity (Rover), Canadarm, the da Vinci Surgical System and Sensei robotic catheter system. These mechanical marvels have revolutionized space exploration and surgery. The chapter explores these innovations and their design.

Mars Exploration Rover

Artist's conception of rover on Mars
Part of a panorama taken by the *Spirit* in May 2004

NASA's Mars Exploration Rover Mission (MER) is an ongoing robotic space mission involving two rovers, *Spirit* and *Opportunity*, exploring the planet Mars. It began in 2003 with the sending of the two rovers—MER-A *Spirit* and MER-B *Opportunity*—to explore the Martian surface and geology.

Objectives

The mission's scientific objective was to search for and characterize a wide range of rocks and soils that hold clues to past water activity on Mars. The mission is part of

NASA's Mars Exploration Program, which includes three previous successful landers: the two Viking program landers in 1976 and Mars Pathfinder probe in 1997.

The total cost of building, launching, landing and operating the rovers on the surface for the initial 90-sol primary mission was US$820 million. Since the rovers have continued to function beyond their initial 90 sol primary mission, they have each received five mission extensions. The fifth mission extension was granted in October 2007, and ran to the end of 2009. The total cost of the first four mission extensions was $104 million, and the fifth mission extension is expected to cost at least $20 million.

In July 2007, during the fourth mission extension, Martian dust storms blocked sunlight to the rovers and threatened the ability of the craft to gather energy through their solar panels, causing engineers to fear that one or both of them might be permanently disabled. However, the dust storms lifted, allowing them to resume operations.

On May 1, 2009, during its fifth mission extension, *Spirit* became stuck in soft soil on Mars. After nearly nine months of attempts to get the rover back on track, including using test rovers on Earth, NASA announced on January 26, 2010 that *Spirit* was being retasked as a stationary science platform. This mode would enable *Spirit* to assist scientists in ways that a mobile platform could not, such as detecting "wobbles" in the planet's rotation that would indicate a liquid core. Jet Propulsion Laboratory (JPL) lost contact with Spirit after last hearing from the rover on March 22, 2010 and continued attempts to regain communications lasted until May 25, 2011, bringing the elapsed mission time to 6 years 2 months 19 days, or over 25 times the original planned mission duration.

In recognition of the vast amount of scientific information amassed by both rovers, two asteroids have been named in their honor: 37452 Spirit and 39382 Opportunity. The mission is managed for NASA by the Jet Propulsion Laboratory, which designed, built, and is operating the rovers.

On January 24, 2014, NASA reported that current studies by the *Curiosity* and *Opportunity* rovers will now be searching for evidence of ancient life, including a biosphere based on autotrophic, chemotrophic and/or chemolithoautotrophic microorganisms, as well as ancient water, including fluvio-lacustrine environments (plains related to ancient rivers or lakes) that may have been habitable. The search for evidence of habitability, taphonomy (related to fossils), and organic carbon on the planet Mars is now a primary NASA objective.

The scientific objectives of the Mars Exploration Rover mission are to:

- Search for and characterize a variety of rocks and soils that hold clues to past water activity. In particular, samples sought include those that have minerals deposited by water-related processes such as precipitation, evaporation, sedimentary cementation, or hydrothermal activity.

- Determine the distribution and composition of minerals, rocks, and soils surrounding the landing sites.

- Determine what geologic processes have shaped the local terrain and influenced the chemistry. Such processes could include water or wind erosion, sedimentation, hydrothermal mechanisms, volcanism, and cratering.

- Perform calibration and validation of surface observations made by Mars Reconnaissance Orbiter instruments. This will help determine the accuracy and effectiveness of various instruments that survey Martian geology from orbit.

- Search for iron-containing minerals, and to identify and quantify relative amounts of specific mineral types that contain water or were formed in water, such as iron-bearing carbonates.

- Characterize the mineralogy and textures of rocks and soils to determine the processes that created them.

- Search for geological clues to the environmental conditions that existed when liquid water was present.

- Assess whether those environments were conducive to life.

During the last two decades, NASA will conduct several missions to address whether life ever existed on Mars. The search begins with determining whether the Martian environment was ever suitable for life. Life, as humans understand it, requires water, hence the history of water on Mars is critical to finding out if the Martian environment was ever conducive to life. Although the Mars Exploration Rovers do not have the ability to detect life directly, they offer important information on the habitability of the environment in the planet's history

NASA's Mars Exploration Rover *Spirit* casts a shadow over the trench that the rover is examining with tools on its robotic arm. *Spirit* took this image with its front hazard-avoidance camera on February 21, 2004, during the rover's 48th martian day, or sol 48.

Opportunity's discarded heat shield

History

The MER-A and MER-B probes were launched on June 10, 2003 and July 7, 2003, respectively. Though both probes launched on Boeing Delta II 7925-9.5 rockets from Cape Canaveral Air Force Station Space Launch Complex 17 (CCAFS SLC-17), MER-B was on the heavy version of that launch vehicle, needing the extra energy for Trans-Mars injection. The launch vehicles were integrated onto pads right next to each other, with MER-A on CCAFS SLC-17A and MER-B on CCAFS SLC-17B. The dual pads allowed for working the 15- and 21-day planetary launch periods close together; the last possible launch day for MER-A was June 19, 2003 and the first day for MER-B was June 25, 2003. NASA's Launch Services Program managed the launch of both spacecraft.

The probes landed in January 2004 in widely separated equatorial locations on Mars.

On January 21, 2004, the Deep Space Network lost contact with *Spirit*, for reasons originally thought to be related to a thunderstorm over Australia. The rover transmitted a message with no data, but later that day missed another communications session with the Mars Global Surveyor. The next day, JPL received a beep from the rover, indicating that it was in fault mode. On January 23, the flight team succeeded in making the rover send. The fault was believed to have been caused by an error in the rover's flash memory subsystem. The rover did not perform any scientific activities for ten days, while engineers updated its software and ran tests. The problem was corrected by reformatting *Spirit's* flash memory and using a software patch to avoid memory overload; *Opportunity* was also upgraded with the patch as a precaution. *Spirit* returned to full scientific operations by February 5.

On March 23, 2004, a news conference was held announcing "major discoveries" of evidence of past liquid water on the Martian surface. A delegation of scientists showed pictures and data revealing a stratified pattern and cross bedding in the rocks of the outcrop inside a crater in Meridiani Planum, landing site of MER-B, *Opportunity*. This suggested that water once flowed in the region. The irregular distribution of chlorine and bromine also suggests that the place was once the shoreline of a salty sea, now evaporated.

On April 8, 2004, NASA announced that it was extending the mission life of the rovers from three to eight months. It immediately provided additional funding of US $15 million through September, and $2.8 million per month for continuing operations. Later that month, *Opportunity* arrived at Endurance crater, taking about five days to drive the 200 meters. NASA announced on September 22 that it was extending the mission life of the rovers for another six months. *Opportunity* was to leave Endurance crater, visit its discarded heat shield, and proceed to Victoria crater. *Spirit* was to attempt to climb to the top of the Columbia Hills.

With the two rovers still functioning well, NASA later announced another 18-month extension of the mission to September 2006. *Opportunity* was to visit the "Etched Terrain" and *Spirit* was to climb a rocky slope toward the top of Husband Hill. On August 21, 2005, *Spirit* reached the summit of Husband Hill after 581 sols and a journey of 4.81 kilometers (2.99 mi).

Spirit's "postcard" view from the summit of Husband Hill: a windswept plateau strewn with rocks, small exposures of outcrop, and sand dunes. The view is to the north, looking down upon the "Tennessee Valley". This approximate true-color composite spans about 90 degrees and consists of eighteen frames captured by the rover's panoramic camera.

Spirit celebrated its one Martian year anniversary (669 sols or 687 Earth days) on November 20, 2005. *Opportunity* celebrated its anniversary on December 12, 2005. At the beginning of the mission, it was expected that the rovers would not survive much longer than 90 Martian days. The Columbia Hills were "just a dream", according to rover driver Chris Leger. *Spirit* explored the semicircular rock formation known as Home Plate. It is a layered rock outcrop that puzzles and excites scientists. It is thought that its rocks are explosive volcanic deposits, though other possibilities exist, including impact deposits or sediment borne by wind or water.

Spirit's front right wheel ceased working on March 13, 2006, while the rover was moving itself to McCool Hill. Its drivers attempted to drag the dead wheel behind Spirit, but this only worked until reaching an impassable sandy area on the lower slopes. Drivers directed *Spirit* to a smaller sloped feature, dubbed "Low Ridge Haven", where it spent the long Martian winter, waiting for spring and increased solar power levels suitable for driving. That September, *Opportunity* reached the rim of Victoria crater, and Spaceflight Now reported that NASA had extended mission for the two rovers through September 2007. On February 6, 2007, *Opportunity* became the first spacecraft to traverse ten kilometers (6.21 miles) on the surface of Mars.

MSL mockup compared with the Mars Exploration Rover and Sojourner rover
by the Jet Propulsion Laboratory on May 12, 2008

Opportunity was poised to enter Victoria Crater from its perch on the rim of Duck Bay on June 28, 2007, but due to extensive dust storms, it was delayed until the dust had cleared and power returned to safe levels. Two months later, *Spirit* and *Opportunity* resumed driving after hunkering down during raging dust storms that limited solar power to a level that nearly caused the permanent failure of both rovers.

On October 1, 2007, both *Spirit* and *Opportunity* entered their fifth mission extension that extended operations into 2009, allowing the rovers to have spent five years exploring the Martian surface, pending their continued survival.

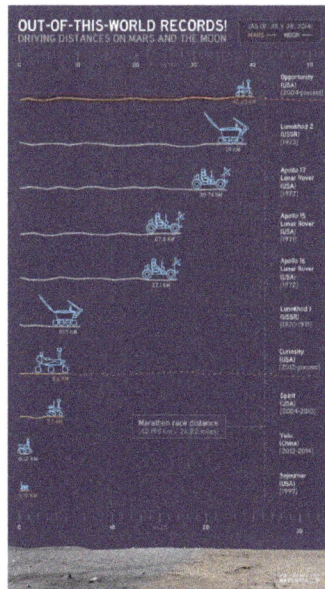

Comparison of distances driven by various wheeled vehicles on
the surface of Earth's moon and Mars (July 28, 2014).

On August 26, 2008, *Opportunity* began its three-day climb out of Victoria crater amidst concerns that power spikes, similar to those seen on *Spirit* before the failure of its right-front wheel, might prevent it from ever being able to leave the crater if a wheel failed. Project scientist Bruce Banerdt also said, "We've done everything we entered Victoria Crater to do and more." *Opportunity* will return to the plains in order to characterize Meridiani Planum's vast diversity of rocks—some of which may have been blasted out of craters such as Victoria. The rover had been exploring Victoria Crater since September 11, 2007. As of January 2009, the two rovers had collectively sent back 250,000 images and traveled over 21 kilometers (13 mi).

After driving about 3.2 kilometers (2.0 mi) since it left Victoria crater, *Opportunity* first saw the rim of Endeavour crater on March 7, 2009. It passed the 10-mile mark (16 kilometers) along the way on sol 1897. Meanwhile, at Gusev crater, *Spirit* was dug in deep into the Martian sand, much as *Opportunity* was at Purgatory Dune in 2005.

On January 3 and January 24, 2010, *Spirit* and *Opportunity* marked six years on Mars, respectively. On January 26, NASA announced that *Spirit* will be used as a stationary research platform after several months of unsuccessful attempts to free the rover from soft sand.

NASA announced on March 24, 2010, that *Opportunity*, which has an estimated remaining drive distance of 12 km to Endeavour Crater, has traveled over 20 km since the start of its mission. Each rover was designed with a mission driving distance goal of just 600 meters. One week later, they announced that *Spirit* may have gone into hibernation for the Martian winter and might not wake up again for months.

On September 8, 2010, it was announced that *Opportunity* had reached the halfway point of the 19-kilometer journey between Victoria crater and Endeavour crater.

On May 22, 2011, NASA announced that it will cease attempts to contact *Spirit*, which has been stuck in a sand trap for two years. The last successful communication with the rover was on March 22, 2010. The final transmission to the rover was on May 25, 2011.

In April 2013, a photo sent back by one of the rovers became widely circulated on social networking and news sites such as Reddit that appeared to depict a human penis carved into the Martian dirt.

On May 16, 2013, NASA announced that *Opportunity* had driven further than any other NASA vehicle on a world other than Earth. After *Opportunity's* total odometry went over 35.744 km (22.210 mi), the rover surpassed the total distance driven by the Apollo 17 Lunar Roving Vehicle.

On July 28, 2014, NASA announced that *Opportunity* had driven further than any other vehicle on a world other than Earth. *Opportunity* covered over 40 km (25 mi), surpassing the total distance of 39 km (24 mi) driven by the Lunokhod 2 lunar rover, the previous record-holder.

On March 23, 2015, NASA announced that *Opportunity* had driven the full 26.2-mile distance of a marathon, with a finish time of roughly 11 years, 2 months.

Spacecraft Design

Delta II lifting off with MER-A on June 10, 2003

Delta II Heavy (7925H-9.5) lifting off from pad 17-B carrying MER-B

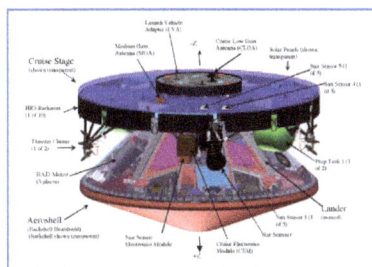

MER cruise stage diagram (Courtesy NASA/JPL-Caltech)

Cruise stage of *Opportunity* rover

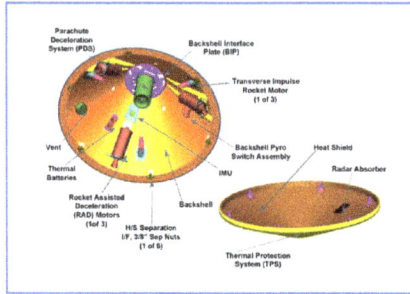

Overview of the Mars Exploration Rover aeroshell

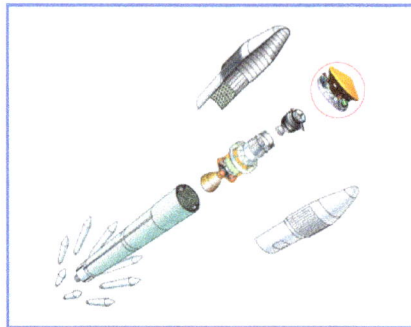

MER launch configuration, break apart illustration

The Mars Exploration Rover was designed to be stowed in the nose of a Delta II rocket. Each spacecraft consists of several components:

- Rover: 185 kg (408 lb)

- Lander: 348 kg (767 lb)

- Backshell / Parachute: 209 kg (461 lb)

- Heat Shield: 78 kg (172 lb)

- Cruise Stage: 193 kg (425 lb)

- Propellant: 50 kg (110 lb)

- Instruments: 5 kg (11 lb)

Total mass is 1,063 kg (2,343 lb).

Cruise Stage

The cruise stage is the component of the spacecraft that is used for travel from Earth to Mars. It is very similar to the Mars Pathfinder in design and is approximately 2.65 meters (8.7 ft) in diameter and 1.6 m (5.2 ft) tall, including the entry vehicle.

The primary structure is aluminium with an outer ring of ribs covered by the solar panels, which are about 2.65 m (8.7 ft) in diameter. Divided into five sections, the solar arrays can provide up to 600 watts of power near Earth and 300 W at Mars.

Heaters and multi-layer insulation keep the electronics "warm". A freon system removes heat from the flight computer and communications hardware inside the rover so they do not overheat. Cruise avionics systems allow the flight computer to interface with other electronics, such as the sun sensors, star scanner and heaters.

Navigation

The star scanner (without a backup system) and sun sensor allowed the spacecraft to know its orientation in space by analyzing the position of the Sun and other stars in relation to itself. Sometimes the craft could be slightly off course; this was expected, given the 500 million kilometer (320 million mile) journey. Thus navigators planned up to six trajectory correction maneuvers, along with health checks.

To ensure the spacecraft arrived at Mars in the right place for its landing, two light-weight, aluminium-lined tanks carried about 31 kg (about 68 lb) of hydrazine propellant. Along with cruise guidance and control systems, the propellant allowed navigators to keep the spacecraft on course. Burns and pulse firings of the propellant allowed three types of maneuvers:

- An axial burn uses pairs of thrusters to change spacecraft velocity;

- A lateral burn uses two "thruster clusters" (four thrusters per cluster) to move the spacecraft "sideways" through seconds-long pulses;

- Pulse mode firing uses coupled thruster pairs for spacecraft precession maneuvers (turns).

Communication

The spacecraft used a high-frequency X band radio wavelength to communicate, which allowed for less power and smaller antennas than many older craft, which used S band.

Navigators sent commands through two antennas on the cruise stage: a cruise low-gain antenna mounted inside the inner ring, and a cruise medium-gain antenna in the outer ring. The low-gain antenna was used close to Earth. It is omni-directional, so the transmission power that reached Earth fell faster with increasing distance. As the craft moved closer to Mars, the Sun and Earth moved closer in the sky as viewed from the craft, so less energy reached Earth. The spacecraft then switched to the medium-gain antenna, which directed the same amount of transmission power into a tighter beam toward Earth.

During flight, the spacecraft was spin-stabilized with a spin rate of two revolutions per minute (rpm). Periodic updates kept antennas pointed toward Earth and solar panels toward the Sun.

Aeroshell

The aeroshell maintained a protective covering for the lander during the seven-month voyage to Mars. Together with the lander and the rover, it constituted the "entry vehicle". Its main purpose was to protect the lander and the rover inside it from the intense heat of entry into the thin Martian atmosphere. It was based on the Mars Pathfinder and Mars Viking designs.

Parts

The aeroshell was made of two main parts: a heat shield and a backshell. The heat shield was flat and brownish, and protected the lander and rover during entry into the Martian atmosphere and acted as the first aerobrake for the spacecraft. The backshell was large, cone-shaped and painted white. It carried the parachute and several components used in later stages of entry, descent, and landing, including:

- A parachute (stowed at the bottom of the backshell);

- The backshell electronics and batteries that fire off pyrotechnic devices like separation nuts, rockets and the parachute mortar;

- A Litton LN-200 Inertial Measurement Unit (IMU), which monitors and reports the orientation of the backshell as it swings under the parachute;

- Three large solid rocket motors called RAD rockets (Rocket Assisted Descent), each providing about a ton of force (10 kilonewtons) for about 60 seconds;

- Three small solid rockets called TIRS (mounted so that they aim horizontally out the sides of the backshell) that provide a small horizontal kick to the backshell to help orient the backshell more vertically during the main RAD rocket burn.

Composition

Built by the Lockheed Martin Astronautics Co. in Denver, Colorado, the aeroshell is made of an aluminium honeycomb structure sandwiched between graphite-epoxy face sheets. The outside of the aeroshell is covered with a layer of phenolic honeycomb. This honeycomb is filled with an ablative material (also called an "ablator"), that dissipates heat generated by atmospheric friction.

The ablator itself is a unique blend of cork wood, binder and many tiny silica glass spheres. It was invented for the heat shields flown on the Viking Mars lander missions. A similar technology was used in the first US manned space missions Mercury, Gemini and Apollo. It was specially formulated to react chemically with the Martian atmosphere during entry and carry heat away, leaving a hot wake of gas behind the vehicle. The vehicle slowed from 19000 km/h (about 12000 mph) to about 1600 km/h

(1000 mph) in about a minute, producing about 60 m/s² (6 g) of acceleration on the lander and rover.

The backshell and heat shield are made of the same materials, but the heat shield has a thicker, $\frac{1}{2}$-inch (13 mm), layer of the ablator. Instead of being painted, the backshell was covered with a very thin aluminized PET film blanket to protect it from the cold of deep space. The blanket vaporized during entry into the Martian atmosphere.

Parachute

Mars Exploration Rover's parachute test

The parachute helped slow the spacecraft during entry, descent, and landing. It is located in the backshell.

Design

The 2003 parachute design was part of a long-term Mars parachute technology development effort and is based on the designs and experience of the Viking and Pathfinder missions. The parachute for this mission is 40% larger than Pathfinder's because the largest load for the Mars Exploration Rover is 80 to 85 kilonewtons (kN) or 18,000 to 19,000 lbf (85 kN) when the parachute fully inflates. By comparison, Pathfinder's inflation loads were approximately 35 kN (about 8,000 lbf). The parachute was designed and constructed in South Windsor, Connecticut by Pioneer Aerospace, the company that also designed the parachute for the *Stardust* mission.

Composition

The parachute is made of two durable, lightweight fabrics: polyester and nylon. A triple bridle made of Kevlar connects the parachute to the backshell.

The amount of space available on the spacecraft for the parachute is so small that the parachute had to be pressure-packed. Before launch, a team tightly folded the 48 suspension lines, three bridle lines, and the parachute. The parachute team loaded the parachute in a special structure that then applied a heavy weight to the parachute package several times. Before placing the parachute into the backshell, the parachute was heat set to sterilize it.

Connected Systems

Descent is halted by retrorockets and lander is dropped 10m (30 ft) to the surface in this computer generated impression.

Zylon Bridles: After the parachute was deployed at an altitude of about 10 km (6.2 mi) above the surface, the heatshield was released using 6 separation nuts and push-off springs. The lander then separated from the backshell and "rappelled" down a metal tape on a centrifugal braking system built into one of the lander petals. The slow descent down the metal tape placed the lander in position at the end of another bridle (tether), made of a nearly 20 m (65 ft) long braided Zylon.

Zylon is an advanced fiber material, similar to Kevlar, that is sewn in a webbing pattern (like shoelace material) to make it stronger. The Zylon bridle provides space for airbag deployment, distance from the solid rocket motor exhaust stream, and increased stability. The bridle incorporates an electrical harness that allows the firing of the solid rockets from the backshell as well as provides data from the backshell inertial measurement unit (which measures rate and tilt of the spacecraft) to the flight computer in the rover.

Rocket assisted descent (RAD) motors: Because the atmospheric density of Mars is less than 1% of Earth's, the parachute alone could not slow down the Mars Exploration Rover enough to ensure a safe, low landing speed. The spacecraft descent was assisted by rockets that brought the spacecraft to a dead stop 10–15 m (30–50 ft) above the Martian surface.

Radar altimeter unit: A radar altimeter unit was used to determine the distance to the Martian surface. The radar's antenna is mounted at one of the lower corners of the lander tetrahedron. When the radar measurement showed the lander was the correct distance above the surface, the Zylon bridle was cut, releasing the lander from the parachute and backshell so that it was free and clear for landing. The radar data also enabled the timing sequence on airbag inflation and backshell RAD rocket firing.

Airbags

Artist's concept of inflated airbags

Airbags used in the Mars Exploration Rover mission are the same type that Mars Pathfinder used in 1997. They had to be strong enough to cushion the spacecraft if it landed on rocks or rough terrain and allow it to bounce across Mars' surface at highway speeds (about 100 km/h) after landing. The airbags had to be inflated seconds before touchdown and deflated once safely on the ground.

The airbags were made of Vectran, like those on Pathfinder. Vectran has almost twice the strength of other synthetic materials, such as Kevlar, and performs better in cold temperatures. Six 100 denier (10 mg/m) layers of Vectran protected one or two inner bladders of Vectran in 200 denier (20 mg/m). Using 100 denier (10 mg/m) leaves more fabric in the outer layers where it is needed, because there are more threads in the weave.

Each rover used four airbags with six lobes each, all of which were connected. Connection was important, since it helped abate some of the landing forces by keeping the bag system flexible and responsive to ground pressure. The airbags were not attached directly to the rover, but were held to it by ropes crisscrossing the bag structure. The ropes gave the bags shape, making inflation easier. While in flight, the bags were stowed along with three gas generators that are used for inflation.

Lander

The spacecraft lander is a protective shell that houses the rover, and together with the airbags, protects it from the forces of impact.

The lander is a tetrahedron shape, whose sides open like petals. It is strong and light, and made of beams and sheets. The beams consist of layers of graphite fiber woven into a fabric that is lighter than aluminium and more rigid than steel. Titanium fittings are glued and fitted onto the beams to allow it to be bolted together. The rover was held inside the lander by bolts and special nuts that were released after landing with small explosives.

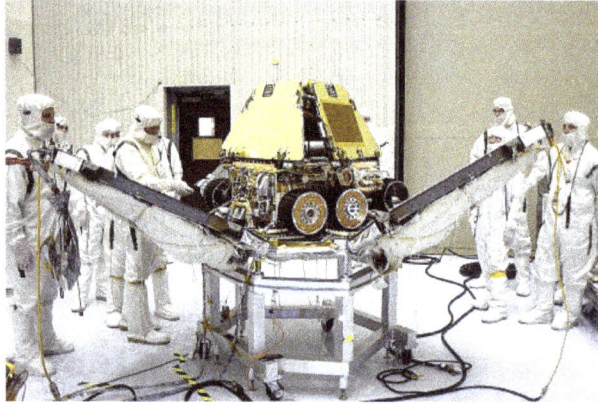

MER lander petals opening (Courtesy NASA/JPL-Caltech)

Uprighting

After the lander stopped bouncing and rolling on the ground, it came to rest on the base of the tetrahedron or one of its sides. The sides then opened to make the base horizontal and the rover upright. The sides are connected to the base by hinges, each of which has a motor strong enough to lift the lander. The rover plus lander has a mass of about 533 kilograms (1,175 pounds). The rover alone has a mass of about 185 kg (408 lb). The gravity on Mars is about 38% of Earth's, so the motor does not need to be as powerful as it would on Earth.

The rover contains accelerometers to detect which way is down (toward the surface of Mars) by measuring the pull of gravity. The rover computer then commanded the correct lander petal to open to place the rover upright. Once the base petal was down and the rover was upright, the other two petals were opened.

The petals initially opened to an equally flat position, so all sides of the lander were straight and level. The petal motors are strong enough so that if two of the petals come to rest on rocks, the base with the rover would be held in place like a bridge above the ground. The base will hold at a level even with the height of the petals resting on rocks, making a straight flat surface throughout the length of the open, flattened lander. The flight team on Earth could then send commands to the rover to adjust the petals and create a safe path for the rover to drive off the lander and onto the Martian surface without dropping off a steep rock.

Moving the Payload onto Mars

The moving of the rover off the lander is called the egress phase of the mission. The rover must avoid having its wheels caught in the airbag material or falling off a sharp incline. To help this, a retraction system on the petals slowly drags the airbags toward the lander before the petals open. Small ramps on the petals fan out to fill spaces between the petals. They cover uneven terrain, rock obstacles, and airbag material, and form a circular area from which the rover can drive off in more directions. They also lower the step that the rover must climb down. They are nicknamed "batwings", and are made of Vectran cloth.

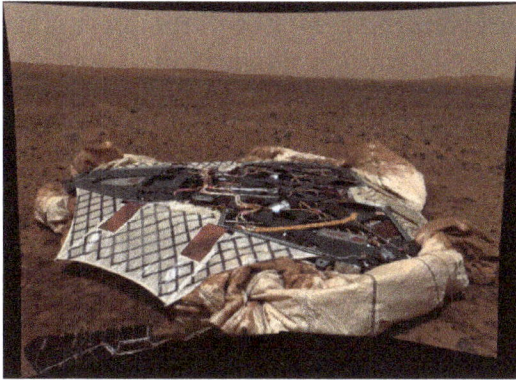

Spirit's lander on Mars

About three hours were allotted to retract the airbags and deploy the lander petals.

Rover Design

Mars Exploration Rover (rear) vs. Sojourner rover (Courtesy NASA/JPL-Caltech)

The rovers are six-wheeled, solar-powered robots that stand 1.5 m (4.9 ft) high, 2.3 m (7.5 ft) wide and 1.6 m (5.2 ft) long. They weigh 180 kg (400 lb), 35 kg (80 lb) of which is the wheel and suspension system.

Drive System

Each rover has six wheels mounted on a rocker-bogie suspension system that ensures wheels remain on the ground while driving over rough terrain. The design reduces the range of motion of the rover body by half, and allows the rover to go over obstacles or through holes that are more than a wheel diameter (250 millimeters (9.8 in)) in size. Each wheel also has cleats, providing grip for climbing in soft sand and scrambling over rocks.

Each wheel has its own motor. The two front and two rear wheels each have individual steering motors. This allows the vehicle to turn in place, a full revolution, and to swerve and curve, making arching turns. The rover is designed to withstand a tilt of 45 degrees in any direction without overturning. However, the rover is programmed through its "fault protection limits" in its hazard avoidance software to avoid exceeding tilts of 30 degrees.

Each rover can spin one of its front wheels in place to grind deep into the terrain. It is to remain motionless while the digging wheel is spinning. The rovers have a top speed on flat hard ground of 50 mm/s (2 in/s). The average speed is 10 mm/s, because its hazard avoidance software causes it to stop every 10 seconds for 20 seconds to observe and understand the terrain into which it has driven.

Power and Electronic Systems

Circular projection showing MER-A *Spirit*'s solar panels covered in dust in October 2007 on Mars. Unexpected cleaning events have periodically increased power.

When fully illuminated, the rover triplejunction solar arrays generate about 140 watts for up to four hours per Martian day (sol). The rover needs about 100 watts to drive. Its power system includes two rechargeable lithium ion batteries weighing 7.15 kg (16 pounds) each, that provide energy when the sun is not shining, especially at night. Over time, the batteries will degrade and will not be able to recharge to full capacity.

For comparison, the Mars Science Laboratory's power system is composed of a Multi-Mission Radioisotope Thermoelectric Generator (MMRTG) produced by Boeing. The MMRTG is designed to provide 125W of electrical power at the start of the mission, falling to 100W after 14 years of service. It is used to power the MSL's many systems and instruments. Solar panels were also considered for the MSL, but RTGs provide constant power, regardless of the time of day, and thus the versatility to work in dark environments and high latitudes where solar energy is not readily available. The MSL generates 2.5 kilowatt hours per day, compared to the Mars Exploration Rovers, which can generate about 0.6 kilowatt hours per day.

It was thought that by the end of the 90-sol mission, the capability of the solar arrays to generate power would likely be reduced to about 50 watts. This was due to antici-pated dust coverage on the solar arrays, and the change in season. Over three Earth years later, however, the rovers' power supplies hovered between 300 watt-hours and 900 watt-hours per day, depending on dust coverage. Cleaning events (dust removal by wind) have occurred more often than NASA expected, keeping the arrays relatively free of dust and extending the life of the mission. During a 2007 global dust storm on Mars, both rovers experienced some of the lowest power of the mission; *Opportunity* dipped to 128 watt-hours. In November 2008, Spirit had overtaken this low-energy record with a production of 89 watt-hours, due to dust storms in the region of Gusev crater.

The rovers run a VxWorks embedded operating system on a radiation-hardened 20 MHz RAD6000 CPU with 128 MB of DRAM with error detection and correction and 3 MB of EEPROM. Each rover also has 256 MB of flash memory. To survive during the various mission phases, the rover's vital instruments must stay within a temperature of −40 °C to +40 °C (−40 °F to 104 °F). At night, the rovers are heated by eight radio-isotope heater units (RHU), which each continuously generate 1 W of thermal energy from the decay of radioisotopes, along with electrical heaters that operate only when necessary. A sputtered gold film and a layer of silica aerogel are used for insulation.

Communication

Pancam Mast Assembly (PMA)

The rover has an X band low-gain and an X band high-gain antenna for communica-tions to and from the Earth, as well as an ultra high frequency monopole antenna for relay communications. The low-gain antenna is omnidirectional, and transmits data at a low rate to Deep Space Network (DSN) antennas on Earth. The high-gain antenna is directional and steerable, and can transmit data to Earth at a higher rate. The rovers use the UHF monopole and its CE505 radio to communicate with spacecraft orbiting Mars, the Mars Odyssey and (before its failure) the Mars Global Surveyor (already more than 7.6 terabits of data were transferred using its Mars Relay antenna and Mars Orbiter Camera's memory buffer of 12 MB). Since MRO went into orbit around Mars, the landers have also used it as a relay asset. Most of the lander data is relayed to Earth through Odyssey and MRO. The orbiters can receive rover signals at a much higher data rate than the Deep Space Network can, due to the much shorter distances from rover to orbiter. The orbiters then quickly relay the rover data to the Earth using their large and high-powered antennas.

Rock Abrasion Tool (RAT)

Alpha particle X-ray spectrometer (APXS) (Courtesy NASA/JPL-Caltech)

Each rover has a total of 9 cameras, which produce 1024-pixel by 1024-pixel images at 12 bits per pixel, but most navigation camera images and image thumbnails are truncated to 8 bits per pixel to conserve memory and transmission time. All images are then compressed using ICER before being stored and sent to Earth. Navigation, thumbnail, and many other image types are compressed to approximately 0.8 to 1.1 bits/pixel. Lower bit rates (less than 0.5 bit/pixel) are used for certain wavelengths of multi-color panoramic images.

ICER is based on wavelets, and was designed specifically for deep-space applications. It produces progressive compression, both lossless and lossy, and incorporates an error-containment scheme to limit the effects of data loss on the deep-space channel. It outperforms the lossy JPEG image compressor and the lossless Rice compressor used by the Mars Pathfinder mission.

Scientific Instrumentation

The rover has various instruments. Three are mounted on the Pancam Mast Assembly (PMA):

- Panoramic Cameras (Pancam), two cameras with color filter wheels for determining the texture, color, mineralogy, and structure of the local terrain.

- Navigation Cameras (Navcam), two cameras that have larger fields of view but lower resolution and is monochromatic, for navigation and driving.

- A periscope assembly for the Miniature Thermal Emission Spectrometer (Mini-TES), which identifies promising rocks and soils for closer examination, and determines the processes that formed them. The Mini-TES was built by Arizona State University. The periscope assembly features two beryllium fold mirrors, a shroud that closes to minimize dust contamination in the assembly, and stray-light rejection baffles that are strategically placed within the graphite epoxy tubes.

The cameras are mounted 1.5 meters high on the Pancam Mast Assembly. The PMA is deployed via the Mast Deployment Drive (MDD). The Azimuth Drive, mounted directly above the MDD, turns the assembly horizontally a whole revolution with signals transmitted through a rolling tape configuration. The camera drive points the cameras in elevation, almost straight up or down. A third motor points the Mini-TES fold mirrors and protective shroud, up to 30° above the horizon and 50° below. The PMA's conceptual design was done by Jason Suchman at JPL, the Cognizant Engineer who later served as Contract Technical Manager (CTM) once the assembly was built by Ball Aerospace & Technologies Corp., Boulder, Colorado. Raul Romero served as CTM once subsystem-level testing began. Satish Krishnan did the conceptual design of the High-Gain Antenna Gimbal (HGAG), whose detailed design, assembly, and test was also performed by Ball Aerospace & Technologies Corp., Boulder, Colorado at which point Satish acted as the CTM.

Four monochromatic hazard cameras (Hazcams) are mounted on the rover's body, two in front and two behind.

The instrument deployment device (IDD), also called the rover arm, holds the following:

- Mössbauer spectrometer (MB) MIMOS II, developed by Dr. Göstar Klingelhöfer at the Johannes Gutenberg University in Mainz, Germany, is used for close-up investigations of the mineralogy of iron-bearing rocks and soils.

- Alpha particle X-ray spectrometer (APXS), developed by the Max Planck Institute for Chemistry in Mainz, Germany, is used for close-up analysis of the abundances of elements that make up rocks and soils. Universities involved in developing the APXS include the University of Guelph, University of California, and Cornell University

- Magnets, for collecting magnetic dust particles, developed by Jens Martin Knudsen's group at the Niels Bohr Institute, Copenhagen. The particles are analyzed by the Mössbauer Spectrometer and X-ray Spectrometer to help determine the ratio of magnetic particles to non-magnetic particles and the composition of magnetic minerals in airborne dust and rocks that have been ground by the Rock Abrasion Tool. There are also magnets on the front of the rover, which are studied extensively by the Mössbauer spectrometer.

- Microscopic Imager (MI) for obtaining close-up, high-resolution images of rocks and soils. Development was led by Ken Herkenhoff's team at the USGS Astrogeology Research Program.

- Rock Abrasion Tool (RAT), developed by Honeybee Robotics, for removing dusty and weathered rock surfaces and exposing fresh material for examination by instruments on board.

The robotic arm is able to place instruments directly up against rock and soil targets of interest.

Naming of Spirit and Opportunity

Sofi Collis with a model of Mars Exploration Rover

The *Spirit* and *Opportunity* rovers were named through a student essay competition. The winning entry was by Sofi Collis, a third-grade Russian-American student from Arizona.

I used to live in an orphanage. It was dark and cold and lonely. At night, I looked up at the sparkly sky and felt better. I dreamed I could fly there. In America, I can make all my dreams come true. Thank you for the 'Spirit' and the 'Opportunity.'

— Sofi Collis, age 9

Prior to this, during the development and building of the rovers, they were known as MER-1 (*Opportunity*) and MER-2 (*Spirit*). Internally, NASA also uses the mission designations MER-A (*Spirit*) and MER-B (*Opportunity*) based on the order of landing on Mars (*Spirit* first then *Opportunity*).

Test Rovers

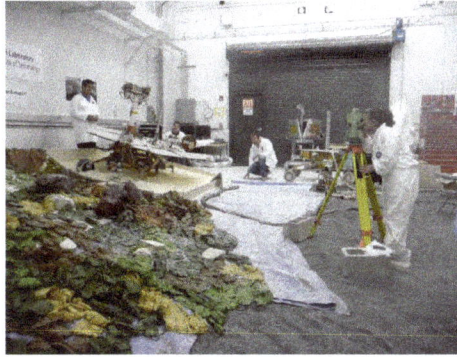

Rover team members simulate *Spirit* in a Martian sandtrap.

The Jet Propulsion Laboratory maintains a pair of rovers, the *Surface System Test-Beds* (SSTB) at its location in Pasadena for testing and modeling of situations on Mars. One test rover, SSTB1, weighing approximately 180 kilograms (400 lb), is fully instrumented and nearly identical to *Spirit* and *Opportunity*. Another test version, *SSTB-Lite*, is identical in size and drive characteristics but does not include all instruments. It weighs in at 80 kilograms (180 lb), much closer to the weight of *Spirit* and *Opportunity* in the reduced gravity of Mars. These rovers were used in 2009 for a simulation of the incident in which *Spirit* became trapped in soft soil.

SAP

The NASA team uses a software application called *SAP* to view images collected from the rover, and to plan its daily activities. There is a version available to the public called Maestro.

Planetary Science Findings

Spirit Landing Site, Gusev Crater

Plains

Although the Gusev crater appears from orbital images to be a dry lakebed, the observations from the surface show the interior plains mostly filled with debris. The rocks on the plains of Gusev are a type of basalt. They contain the minerals olivine, pyroxene, plagioclase, and magnetite, and they look like volcanic basalt as they are fine-grained with irregular holes (geologists would say they have vesicles and vugs). Much of the soil on the plains came from the breakdown of the local rocks. Fairly high levels of nickel were found in some soils; probably from meteorites. Analysis shows that the rocks have been slightly altered by tiny amounts of water. Outside coatings and cracks inside the rocks suggest water deposited minerals, maybe bromine compounds. All the rocks contain a fine coating of dust and one or more harder rinds of material. One type can be brushed off, while another needed to be ground off by the Rock Abrasion Tool (RAT).

There are a variety of rocks in the Columbia Hills (Mars), some of which have been altered by water, but not by very much water.

These rocks can be classified in different ways. The amounts and types of minerals make the rocks primitive basalts—also called picritic basalts. The rocks are similar to ancient terrestrial rocks called basaltic komatiites. Rocks of the plains also resemble the basaltic shergottites, meteorites which came from Mars. One classification system compares the amount of alkali elements to the amount of silica on a graph; in this system, Gusev plains rocks lie near the junction of basalt, picrobasalt, and tephite. The Irvine-Barager classification calls them basalts. Plain's rocks have been very slightly altered, probably by thin films of water because they are softer and contain veins of light colored material that may be bromine compounds, as well as coatings or rinds. It is thought that small amounts of water may have gotten into cracks inducing mineralization processes). Coatings on the rocks may have occurred when rocks were buried and interacted with thin films of water and dust. One sign that they were altered was that it was easier to grind these rocks compared to the same types of rocks found on Earth.

The first rock that *Spirit* studied was Adirondack. It turned out to be typical of the other rocks on the plains.

First color picture from Gusev crater. Rocks were found to be basalt. Everything was covered with a fine dust that *Spirit* determined was magnetic because of the mineral magnetite.

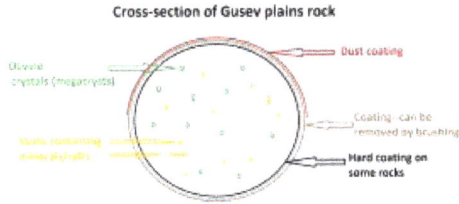

Cross-sectional drawing of a typical rock from the plains of Gusev crater. Most rocks contain a coating of dust and one or more harder coatings. Veins of water-deposited veins are visible, along with crystals of olivine. Veins may contain bromine salts.

Dust

The dust in Gusev Crater is the same as dust all around the planet. All the dust was found to be magnetic. Moreover, *Spirit* found the magnetism was caused by the mineral magnetite, especially magnetite that contained the element titanium. One magnet was able to completely divert all dust hence all Martian dust is thought to be magnetic. The spectra of the dust was similar to spectra of bright, low thermal inertia regions like Tharsis and Arabia that have been detected by orbiting satellites. A thin layer of dust, maybe less than one millimeter thick covers all surfaces. Something in it contains a small amount of chemically bound water.

Columbia Hills

As the rover climbed above the plains onto the Columbia Hills, the mineralogy that was seen changed. Scientists found a variety of rock types in the Columbia Hills, and they placed them into six different categories. The six are: Clovis, Wishbone, Peace, Watchtower, Backstay, and Independence. They are named after a prominent rock in each group. Their chemical compositions, as measured by APXS, are significantly different from each other. Most importantly, all of the rocks in Columbia Hills show various degrees of alteration due to aqueous fluids. They are enriched in the elements phosphorus, sulfur, chlorine, and bromine—all of which can be carried around in water solutions. The Columbia Hills' rocks contain basaltic glass, along with varying amounts of olivine and sulfates. The olivine abundance varies inversely with the amount of sulfates. This is exactly what is expected because water destroys olivine but helps to produce sulfates.

The Clovis group is especially interesting because the Mössbauer spectrometer (MB) detected goethite in it. Goethite forms only in the presence of water, so its discovery is the first direct evidence of past water in the Columbia Hills's rocks. In addition, the MB spectra of rocks and outcrops displayed a strong decline in olivine presence, although the rocks probably once contained much olivine. Olivine is a marker for the lack of water because it easily decomposes in the presence of water. Sulfate was found, and it needs water to form. Wishstone contained a great deal of plagioclase, some olivine, and anhydrate (a sulfate). Peace rocks showed sulfur and strong evidence for bound water, so hydrated sulfates are suspected. Watchtower class rocks lack olivine consequently they may have been altered by water. The Independence class showed some signs of clay (perhaps montmorillonite a member of the smectite group). Clays require fairly long term exposure to water to form. One type of soil, called Paso Robles, from the Columbia Hills, may be an evaporate deposit because it contains large amounts of sulfur, phosphorus, calcium, and iron. Also, MB found that much of the iron in Paso Robles soil was of the oxidized, Fe^{3+} form. Towards the middle of the six-year mission (a mission that was supposed to last only 90 days), large amounts of pure silica were found in the soil. The silica could have come from the interaction of soil with acid vapors produced by volcanic activity in the presence of water or from water in a hot spring environment.

After *Spirit* stopped working scientists studied old data from the Miniature Thermal Emission Spectrometer, or Mini-TES and confirmed the presence of large amounts of carbonate-rich rocks, which means that regions of the planet may have once harbored water. The carbonates were discovered in an outcrop of rocks called "Comanche."

In summary, *Spirit* found evidence of slight weathering on the plains of Gusev, but no evidence that a lake was there. However, in the Columbia Hills there was clear evidence for a moderate amount of aqueous weathering. The evidence included sulfates and the minerals goethite and carbonates which only form in the presence of water. It is believed that Gusev crater may have held a lake long ago, but it has since been covered by igneous materials. All the dust contains a magnetic component which was identified as magnetite with some titanium. Furthermore, the thin coating of dust that covers everything on Mars is the same in all parts of Mars.

Opportunity Landing Site, Meridiani Planum

Self-portrait of *Opportunity* near Endeavour Crater on the surface of Mars (January 6, 2014).

The *Opportunity* rover landed in a small crater, dubbed "Eagle", on the flat plains of Meridiani. The plains of the landing site were characterized by the presence of a large number of small spherules, spherical concretions that were tagged "blueberries" by the science team, which were found both loose on the surface, and also embedded in the rock. These proved to have a high concentration of the mineral hematite, and showed the signature of being formed in an aqueous environment. The layered bedrock revealed in the crater walls showed signs of being sedimentary in nature, and compositional and microscopic-imagery analysis showed this to be primarily with composition of Jarosite, a ferrous sulfate mineral that is characteristically an evaporite that is the residue from the evaporation of a salty pond or sea.

The mission has provided substantial evidence of past water activity on Mars. In addition to investigating the "water hypothesis", *Opportunity* has also obtained astronomical observations and atmospheric data. The extended mission took the rover across the plains to a series of larger craters in the south, with the arrival at the edge of a 25-km diameter crater, Endeavour Crater, eight years after landing. The orbital spectroscopy of this crater rim show the signs of phyllosilicate rocks, indicative of older sedimentary deposits.

Related

- Spirit rover (MER-A) · Opportunity rover (MER-B) · Scientific information from the Mars Exploration Rover mission · Cleaning event

- List of surface features of Mars seen by the Spirit rover · List of surface features of Mars seen by the Opportunity rover

- 37452 Spirit · 39382 Opportunity

Instruments

- APXS · MIMOS II · Mini-TES · Rock Abrasion Tool (RAT) · Hazcam · Pancam · Navcam

Communication

- Low gain antenna · High gain antenna · 2001 Mars Odyssey · Mars Reconnaissance Orbiter · Deep Space Network · Goldstone DSCC · X-Band

Other Systems

- Delta II · Maestro · VxWorks · RAD6000 · Radioisotope heater unit · Rocker-bogie · Multijunction photovoltaic cell · Lithium ion battery

Supporting Institutions

- NASA · JPL · Boeing IDS · Cornell University · Deep Space Network · Arizona State University · The Aerospace Corporation · Ball Aerospace · Johannes Gutenberg University · Max Planck Institute for Chemistry · Niels Bohr Institute · USGS Astrogeology Research Program · Honeybee Robotics

Curiosity (Rover)

Curiosity is a car-sized robotic rover exploring Gale Crater on Mars as part of NASA's Mars Science Laboratory mission (MSL). As of September 23, 2016, *Curiosity* has been on Mars for 1469 sols (1509 total days) since landing on August 6, 2012.

Curiosity was launched from Cape Canaveral on November 26, 2011, at 15:02 UTC aboard the MSL spacecraft and landed on Aeolis Palus in Gale Crater on Mars on August 6, 2012, 05:17 UTC. The Bradbury Landing site was less than 2.4 km (1.5 mi) from the center of the rover's touchdown target after a 563,000,000 km (350,000,000 mi) journey.

The rover's goals include: investigation of the Martian climate and geology; assessment of whether the selected field site inside Gale Crater has ever offered environmental conditions favorable for microbial life, including investigation of the role of water; and planetary habitability studies in preparation for future human exploration.

Curiosity's design will serve as the basis for the planned Mars 2020 rover. In December 2012, *Curiosity*'s two-year mission was extended indefinitely.

Goals and Objectives

As established by the Mars Exploration Program, the main scientific goals of the MSL mission are to help determine whether Mars could ever have supported life, as well as determining the role of water, and to study the climate and geology of Mars. The mission will also help prepare for human exploration. To contribute to these goals, MSL has eight main scientific objectives:

Biological

1. Determine the nature and inventory of organic carbon compounds

2. Investigate the chemical building blocks of life (carbon, hydrogen, nitrogen, oxygen, phosphorus, and sulfur)

3. Identify features that may represent the effects of biological processes (biosignatures and biomolecules)

Geological and Geochemical

4. Investigate the chemical, isotopic, and mineralogical composition of the Martian surface and near-surface geological materials

5. Interpret the processes that have formed and modified rocks and soils

Planetary Process

6. Assess long-timescale (i.e., 4-billion-year) Martian atmospheric evolution processes

7. Determine present state, distribution, and cycling of water and carbon dioxide

Surface Radiation

8. Characterize the broad spectrum of surface radiation, including galactic and cosmic radiation, solar proton events and secondary neutrons. As part of its exploration, it also measured the radiation exposure in the interior of the spacecraft as it traveled to Mars, and it is continuing radiation measurements as it explores the surface of Mars. This data would be important for a future manned mission.

About one year into the surface mission, and having assessed that ancient Mars could have been hospitable to microbial life, the MSL mission objectives evolved to developing predictive models for the preservation process of organic compounds and biomolecules; a branch of paleontology called taphonomy.

Specifications

Curiosity comprised 23 percent of the mass of the 3,893 kg (8,583 lb) *Mars Science Laboratory* (MSL) spacecraft, which had the sole mission of delivering the rover safely across space from Earth to a soft landing on the surface of Mars. The remaining mass of the MSL craft was discarded in the process of carrying out this task.

- Dimensions: *Curiosity* has a mass of 899 kg (1,982 lb) including 80 kg (180 lb) of scientific instruments. The rover is 2.9 m (9.5 ft) long by 2.7 m (8.9 ft) wide by 2.2 m (7.2 ft) in height.

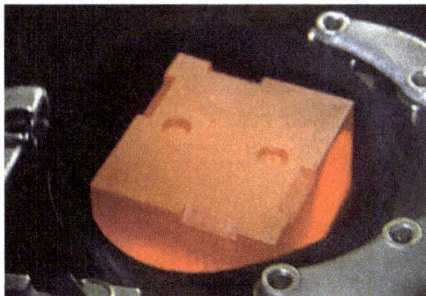

Radioisotope within a graphite shell that goes into the generator

- Power source: *Curiosity* is powered by a radioisotope thermoelectric generator (RTG), like the successful Viking 1 and Viking 2 Mars landers in 1976.

Radioisotope power systems (RPSs) are generators that produce electricity from the decay of radioactive isotopes, such as plutonium-238, which is a non-fissile isotope of plutonium. Heat given off by the decay of this isotope is converted into electric voltage by thermocouples, providing constant power during all seasons and through the day and night. Waste heat can be used via pipes to warm systems, freeing electrical power for the operation of the vehicle and instruments. *Curiosity*'s RTG is fueled by 4.8 kg (11 lb) of plutonium-238 dioxide supplied by the U.S. Department of Energy.

Masthead casts a shadow in this NavCam image on Sol 2 (August 8, 2012)

Curiosity is powered by a Multi-Mission Radioisotope Thermoelectric Generator (MMRTG), designed and built by Rocketdyne and Teledyne Energy Systems under contract to the U.S. Department of Energy, and assembled and tested by the Idaho National Laboratory. Based on legacy RTG technology, it represents a more flexible and compact development step, and is designed to produce 125 watts of electrical power from about 2,000 watts of thermal power at the start of the mission. The MMRTG produces less power over time as its plutonium fuel decays: at its minimum lifetime of 14 years, electrical power output is down to 100 watts. The power source will generate 9 MJ (2.5 kWh) each day, much more than the solar panels of the Mars Exploration Rovers, which can generate about 2.1 MJ (0.58 kWh) each day. The electrical output from the MMRTG charges two rechargeable lithium-ion batteries. This enables the power subsystem to meet peak power demands of rover activities when the demand temporarily exceeds the generator's steady output level. Each battery has a capacity of about 42 ampere-hours.

- Heat rejection system: The temperatures at the landing site can vary from −127 to 40 °C (−197 to 104 °F); therefore, the thermal system will warm the rover for most of the Martian year. The thermal system will do so in several ways: passively, through the dissipation to internal components; by electrical heaters strategically placed on key components; and by using the rover heat rejection system (HRS). It uses fluid pumped through 60 m (200 ft) of tubing in the rover body so that sensitive components are kept at optimal temperatures. The fluid loop serves the additional purpose of rejecting heat when the rover has become too warm, and it can also gather waste heat from the power source by pumping fluid through two heat exchangers that are mounted alongside the RTG. The HRS also has the ability to cool components if necessary.

- Computers: The two identical on-board rover computers, called Rover Computer Element (RCE) contain radiation hardened memory to tolerate the extreme radiation from space and to safeguard against power-off cycles. Each computer's memory includes 256 kB of EEPROM, 256 MB of DRAM, and 2 GB of flash memory. For comparison, the Mars Exploration Rovers used 3 MB of EEPROM, 128 MB of DRAM, and 256 MB of flash memory.

The RCE computers use the RAD750 CPU, which is a successor to the RAD6000 CPU of the Mars Exploration Rovers. The RAD750 CPU, a radiation-hardened version of the PowerPC 750, can execute up to 400 MIPS, while the RAD6000 CPU is capable of up to only 35 MIPS. Of the two on-board computers, one is configured as backup and will take over in the event of problems with the main computer. On February 28, 2013, NASA was forced to switch to the backup computer due to an issue with the then active computer's flash memory, which resulted in the computer continuously rebooting in a loop. The backup computer was turned on in safe mode and subsequently returned to active status on March 4. The same issue happened in late March, resuming full operations on March 25, 2013.

The rover has an Inertial Measurement Unit (IMU) that provides 3-axis information on its position, which is used in rover navigation. The rover's computers are constantly self-monitoring to keep the rover operational, such as by regulating the rover's temperature. Activities such as taking pictures, driving, and operating the instruments are performed in a command sequence that is sent from the flight team to the rover. The rover installed its full surface operations software after the landing because its computers did not have sufficient main memory available during flight. The new software essentially replaced the flight software.

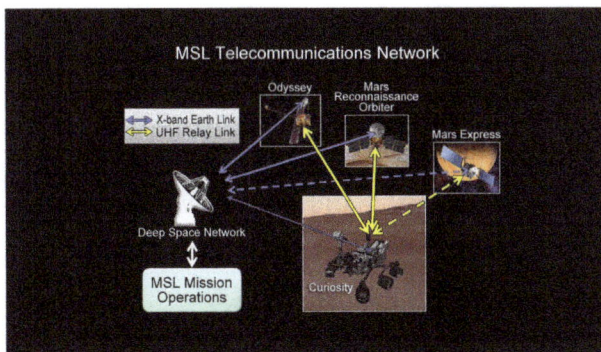

Curiosity transmits to Earth directly or via three relay satellites in Mars orbit.

- Communications: *Curiosity* is equipped with significant telecommunication redundancy by several means – an X band transmitter and receiver that can communicate directly with Earth, and a UHF Electra-Lite software-defined radio for communicating with Mars orbiters. Communication with orbiters is expected to be the main path for data return to Earth, since the orbiters have both more power and larger antennas than the lander allowing for faster transmission speeds. Telecommunication includes a small deep space transponder on the descent stage and a solid-state power amplifier on the rover for X band. The rover also has two UHF radios, the signals of which the 2001 Mars Odyssey satellite is capable of relaying back to Earth. An average of 14 minutes, 6 seconds will be required for signals to travel between Earth and Mars. *Curiosity* can communicate with Earth directly at speeds up to 32 kbit/s, but the bulk of the data transfer should be relayed through the Mars Reconnaissance Orbiter and Odyssey orbiter. Data transfer speeds between *Curiosity* and each orbiter may reach 2 Mbit/s and 256 kbit/s, respectively, but each orbiter is able to communicate with *Curiosity* for only about eight minutes per day (0.56% of the time). Communication from and to *Curiosity* relies on internationally agreed space data communications protocols as defined by the Consultative Committee for Space Data Systems.

  JPL is the central data distribution hub where selected data products are provided to remote science operations sites as needed. JPL is also the central hub for the uplink process, though participants are distributed at their respective home institutions. At landing, telemetry was monitored by three orbiters, depending on their dynamic location: the 2001 Mars Odyssey, Mars Reconnaissance Orbiter and ESA's Mars Express satellite.

- Mobility systems: *Curiosity* is equipped with six 50 cm (20 in) diameter wheels in a rocker-bogie suspension. The suspension system also served as landing gear for the vehicle, unlike its smaller predecessors. Each wheel has cleats and is independently actuated and geared, providing for climbing in soft sand and scrambling over rocks. Each front and rear wheel can be independently steered, allowing the vehicle to turn in place as well as execute arcing turns. Each wheel has a pattern that helps it maintain traction but also leaves patterned tracks in the sandy surface of Mars. That pattern is used by on-board cameras to estimate the distance traveled. The pattern itself is Morse code for "JPL" (·--- ·--· ·--·). The rover is capable of climbing sand dunes with slopes up to 12.5°. Based on the center of mass, the vehicle can withstand a tilt of at least 50° in any direction without overturning, but automatic sensors will limit the rover from exceeding 30° tilts. After two years of use, the wheels are visibly worn with punctures and tears.

  Curiosity can roll over obstacles approaching 65 cm (26 in) in height, and it has a ground clearance of 60 cm (24 in). Based on variables including power levels, terrain difficulty, slippage and visibility, the maximum terrain-traverse speed is estimated to be 200 m (660 ft) per day by automatic navigation. The rover landed about 10 km (6.2 mi) from the base of Mount Sharp, and it is expected to traverse a minimum of 19 km (12 mi) during its primary two-year mission. It can travel up to 90 metres (300 ft) per hour but average speed is about 30 metres (98 ft) per hour.

Instruments

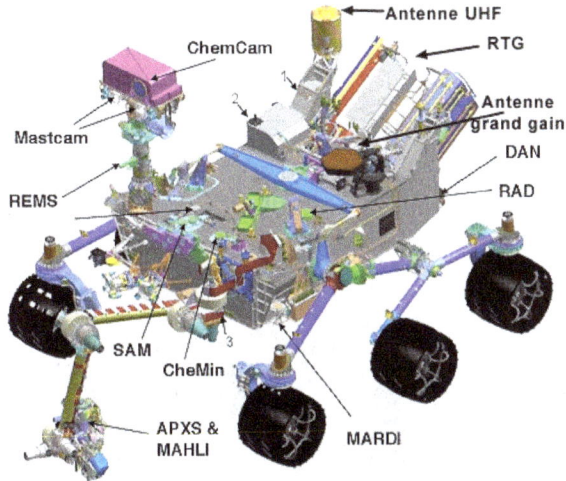

Instrument location diagram

The general sample analysis strategy begins with high-resolution cameras to look for features of interest. If a particular surface is of interest, *Curiosity* can vaporize a small portion of it with an infrared laser and examine the resulting spectra signature to query the rock's elemental composition. If that signature is intriguing, the rover will use its long arm to swing over a microscope and an X-ray spectrometer to take a closer look. If the specimen warrants further analysis, *Curiosity* can drill into the boulder and deliver a powdered sample to either the SAM or the CheMin analytical laboratories inside the rover. The MastCam, Mars Hand Lens Imager (MAHLI), and Mars Descent Imager (MARDI) cameras were developed by Malin Space Science Systems and they all share common design components, such as on-board electronic imaging processing boxes, 1600×1200 CCDs, and an RGB Bayer pattern filter.

It has 17 cameras: HazCams (8), NavCams (4), MastCams (2), MAHLI (1), MARDI (1), and ChemCam (1).

Mast Camera (MastCam)

The MastCam system provides multiple spectra and true-color imaging with two cameras. The cameras can take true-color images at 1600×1200 pixels and up to 10 frames per second hardware-compressed video at 720p (1280×720).

One MastCam camera is the Medium Angle Camera (MAC), which has a 34 mm (1.3 in) focal length, a 15° field of view, and can yield 22 cm/pixel (8.7 in/pixel) scale at 1 km (0.62 mi). The other camera in the MastCam is the Narrow Angle Camera (NAC), which has a 100 mm (3.9 in) focal length, a 5.1° field of view, and can yield 7.4 cm/pixel (2.9 in/pixel) scale at 1 km (0.62 mi). Malin also developed a pair of MastCams with zoom lenses, but these were not included in the rover because of the time required to test the new hardware and the looming November 2011 launch date. However, the improved zoom version was selected to be incorporated on the upcoming Mars 2020 mission as Mastcam-Z.

The turret at the end of the robotic arm holds five devices.

Each camera has eight gigabytes of flash memory, which is capable of storing over 5,500 raw images, and can apply real time lossless data compression. The cameras have an autofocus capability that allows them to focus on objects from 2.1 m (6 ft 11 in) to infinity. In addition to the fixed RGBG Bayer pattern filter, each camera has an eight-position filter wheel. While the Bayer filter reduces visible light throughput, all three colors are mostly transparent at wavelengths longer than 700 nm, and have minimal effect on such infrared observations.

Chemistry and Camera complex (ChemCam)

The internal spectrometer (left) and the laser telescope (right) for the mast

ChemCam is a suite of remote sensing instruments, and as the name implies, Chem-Cam is actually two different instruments combined as one: a laser-induced breakdown spectroscopy (LIBS) and a Remote Micro Imager (RMI) telescope. The ChemCam instrument suite was developed by the French CESR laboratory and the Los Alamos National Laboratory . The flight model of the mast unit was delivered from the French CNES to Los Alamos National Laboratory. The purpose of the LIBS instrument is to provide elemental compositions of rock and soil, while the RMI will give ChemCam scientists high-resolution images of the sampling areas of the rocks and soil that LIBS targets. The LIBS instrument can target a rock or soil sample up to 7 m (23 ft) away, vaporizing a small amount of it with about 50 to 75 5-nanosecond pulses from a 1067 nm infrared laser and then observing the spectrum of the light emitted by the vaporized rock.

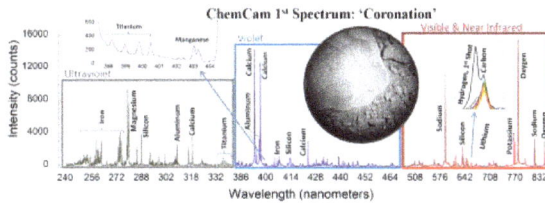

First laser spectrum of chemical elements from ChemCam on *Curiosity*
("Coronation" rock, August 19, 2012)

ChemCam has the ability to record up to 6,144 different wavelengths of ultraviolet, visible, and infrared light. Detection of the ball of luminous plasma will be done in the visible, near-UV and near-infrared ranges, between 240 nm and 800 nm. The first initial laser testing of the ChemCam by *Curiosity* on Mars was performed on a rock, N165 ("Coronation" rock), near Bradbury Landing on August 19, 2012. The ChemCam team expects to take approximately one dozen compositional measurements of rocks per day.

Using the same collection optics, the RMI provides context images of the LIBS analysis spots. The RMI resolves 1 mm (0.039 in) objects at 10 m (33 ft) distance, and has a field of view covering 20 cm (7.9 in) at that distance.

Navigation Cameras (Navcams)

First full-resolution navcam images

Curiosity's self-portrait shows the deck of the rover as viewed from the navcams

The rover has two pairs of black and white navigation cameras mounted on the mast to support ground navigation. The cameras have a 45° angle of view and use visible light to capture stereoscopic 3-D imagery.

Rover Environmental Monitoring Station (REMS)

REMS comprises instruments to measure the Mars environment: humidity, pressure, temperatures, wind speeds, and ultraviolet radiation. It is a meteorological package that includes an ultraviolet sensor provided by the Spanish Ministry of Education and Science. The investigative team is led by Javier Gómez-Elvira of the Center for Astrobiology (Madrid) and includes the Finnish Meteorological Institute as a partner. All sensors are located around three elements: two booms attached to the rover's mast, the Ultraviolet Sensor (UVS) assembly located on the rover top deck, and the Instrument Control Unit (ICU) inside the rover body. REMS will provide new clues about the Martian general circulation, micro scale weather systems, local hydrological cycle, destructive potential of UV radiation, and subsurface habitability based on ground-atmosphere interaction.

Hazard Avoidance Cameras (Hazcams)

The rover has four pairs of black and white navigation cameras called hazcams, two pairs in the front and two pairs in the back. They are used for autonomous hazard avoidance during rover drives and for safe positioning of the robotic arm on rocks and soils. Each camera in a pair is hardlinked to one of two identical main computers for redundancy; only four out of the eight cameras are in use at any one time. The cameras use visible light to capture stereoscopic three-dimensional (3-D) imagery. The cameras have a 120° field of view and map the terrain at up to 3 m (9.8 ft) in front of the rover. This imagery safeguards against the rover crashing into unexpected obstacles, and works in tandem with software that allows the rover to make its own safety choices.

Mars Hand Lens Imager (MAHLI)

Mars Hand Lens Imager (MAHLI) on Mars

Alpha Particle X-Ray Spectrometer (APXS) on Mars

Curiosity's instruments near Bradbury Landing; Mount Sharp is in the background (September 8, 2012).

MAHLI is a camera on the rover's robotic arm, and acquires microscopic images of rock and soil. MAHLI can take true-color images at 1600×1200 pixels with a resolution as high as 14.5 micrometers per pixel. MAHLI has an 18.3 to 21.3 mm (0.72 to 0.84 in) focal length and a 33.8–38.5° field of view. MAHLI has both white and ultraviolet LED illumination for imaging in darkness or fluorescence imaging. MAHLI also has mechanical focusing in a range from infinite to millimetre distances. This system can make some images with focus stacking processing. MAHLI can store either the raw images or do real time lossless predictive or JPEG compression. The calibration target for MAHLI includes color references, a metric bar graphic, a 1909 VDB Lincoln penny, and a stair-step pattern for depth calibration.

Alpha Particle X-ray Spectrometer (APXS)

The device irradiates samples with alpha particles and maps the spectra of X-rays that are re-emitted for determining the elemental composition of samples. *Curiosity*'s APXS was developed by the Canadian Space Agency. MacDonald Dettwiler (MDA), the Canadian aerospace company that built the Canadarm and RADARSAT, were responsible for the engineering design and building of the APXS. The APXS science team includes members from the University of Guelph, the University of New Brunswick, the University of Western Ontario, NASA, the University of California, San Diego and Cornell University. The APXS instrument takes advantage of particle-induced X-ray emission (PIXE) and X-ray fluorescence, previously exploited by the Mars Pathfinder and the Mars Exploration Rovers.

Chemistry and Mineralogy (CheMin)

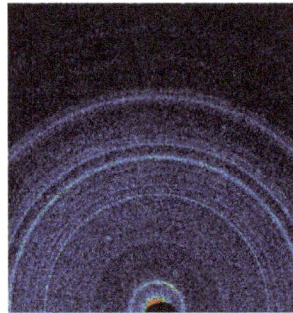

First X-ray diffraction view of Martian soil (*Curiosity* at Rocknest, October 17, 2012).

CheMin is the Chemistry and Mineralogy X-ray powder diffraction and fluorescence instrument. CheMin is one of four spectrometers. It can identify and quantify the abundance of the minerals on Mars. It was developed by David Blake at NASA Ames Research Center and the Jet Propulsion Laboratory, and won the 2013 NASA Government Invention of the year award. The rover can drill samples from rocks and the resulting fine powder is poured into the instrument via a sample inlet tube on the top of the vehicle. A beam of X-rays is then directed at the powder and the crystal structure of the minerals deflects it at characteristic angles, allowing scientists to identify the minerals being analyzed.

On October 17, 2012, at "Rocknest", the first X-ray diffraction analysis of Martian soil was performed. The results revealed the presence of several minerals, including feldspar, pyroxenes and olivine, and suggested that the Martian soil in the sample was similar to the "weathered basaltic soils" of Hawaiian volcanoes. The paragonetic tephra from a Hawaiian cinder cone has been mined to create Martian regolith simulant for researchers to use since 1998.

Sample Analysis at Mars (SAM)

The SAM instrument suite analyzes organics and gases from both atmospheric and solid samples. It consists of instruments developed by the NASA Goddard Space Flight Center, the Laboratoire Inter-Universitaire des Systèmes Atmosphériques (LISA) (jointly operated by France's CNRS and Parisian universities), and Honeybee Robotics, along with many additional external partners. The three main instruments are a Quadrupole Mass Spectrometer (QMS), a gas chromatograph (GC) and a tunable laser spectrometer (TLS). These instruments will perform precision measurements of oxygen and carbon isotope ratios in carbon dioxide (CO_2) and methane (CH_4) in the atmosphere of Mars in order to distinguish between their geochemical or biological origin.

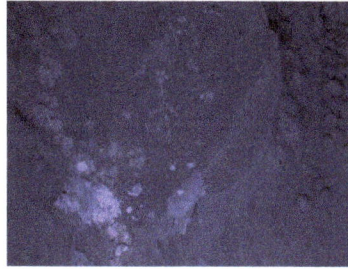

First night-time pictures on Mars (white-light above/UV below)
(*Curiosity* viewing Sayunei rock, January 22, 2013)

Dust Removal Tool (DRT)

First use of *Curiosity*'s Dust Removal Tool (DRT) (January 6, 2013); Ekwir_1 rock before/after cleaning (above) and closeup (below)

The Dust Removal Tool (DRT) is a motorized, wire-bristle brush on the turret at the end of *Curiosity*'s arm. The DRT was first used on a rock target named Ekwir_1 on January 6, 2013. Honeybee Robotics built the DRT.

Radiation Assessment Detector (RAD)

This instrument was the first of ten MSL instruments to be turned on. Its first role was to characterize the broad spectrum of radiation environment found inside the space-craft during the cruise phase. These measurements have never been done before from the inside of a spacecraft in interplanetary space. Its primary purpose is to determine the viability and shielding needs for potential human explorers, as well as to charac-terize the radiation environment on the surface of Mars, which it started doing imme-diately after MSL landed in August 2012. Funded by the Exploration Systems Mission Directorate at NASA Headquarters and Germany's Space Agency (DLR), RAD was de-veloped by Southwest Research Institute (SwRI) and the extraterrestrial physics group at Christian-Albrechts-Universität zu Kiel, Germany.

Dynamic Albedo of Neutrons (DAN)

A pulsed sealed-tube neutron source and detector for measuring hydrogen or ice and water at or near the Martian surface, provided by the Russian Federal Space Agency, and funded by Russia.

Mars Descent Imager (MARDI)

MARDI camera

During the descent to the Martian surface, MARDI took color images at 1600×1200 pix-els with a 1.3-millisecond exposure time starting at distances of about 3.7 km (2.3 mi) to near 5 m (16 ft) from the ground, at a rate of four frames per second for about two minutes. MARDI has a pixel scale of 1.5 m (4.9 ft) at 2 km (1.2 mi) to 1.5 mm (0.059 in) at 2 m (6.6 ft) and has a 90° circular field of view. MARDI has eight gigabytes of inter-nal buffer memory that is capable of storing over 4,000 raw images. MARDI imaging allowed the mapping of surrounding terrain and the location of landing. JunoCam, built for the Juno spacecraft, is based on MARDI.

Robotic Arm

The rover has a 2.1 m (6.9 ft) long robotic arm with a cross-shaped turret holding five devices that can spin through a 350° turning range. The arm makes use of three joints to extend it forward and to stow it again while driving. It has a mass of 30 kg (66 lb) and its diameter, including the tools mounted on it, is about 60 cm (24 in). It was designed, built, and tested by MDA US Systems, building upon their prior robotic arm work on the Mars Surveyor 2001 Lander, the Mars Phoenix (spacecraft), and the two Mars Exploration Rovers, Spirit and Opportunity.

First drill tests (John Klein rock, Yellowknife Bay, February 2, 2013).

Two of the five devices are *in-situ* or contact instruments known as the X-ray spectrometer (APXS), and the Mars Hand Lens Imager (MAHLI camera). The remaining three are associated with sample acquisition and sample preparation functions: a percussion drill; a brush; and mechanisms for scooping, sieving, and portioning samples of powdered rock and soil. The diameter of the hole in a rock after drilling is 1.6 cm (0.63 in) and up to 5 cm (2.0 in) deep. The drill carries two spare bits. The rover's arm and turret system can place the APXS and MAHLI on their respective targets, and also obtain powdered sample from rock interiors, and deliver them to the SAM and CheMin analyzers inside the rover.

Comparisons

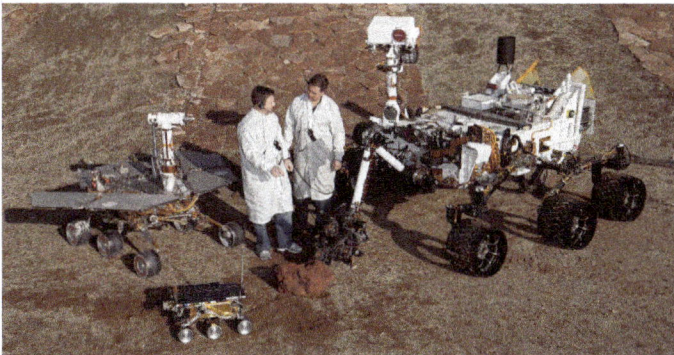

Two Jet Propulsion Laboratory engineers stand with three vehicles, providing a size comparison of three generations of Mars rovers. Front and center is the flight spare for the first Mars rover, *Sojourner*, which landed on Mars in 1997 as part of the Mars

Pathfinder Project. On the left is a Mars Exploration Rover (MER) test vehicle that is a working sibling to *Spirit* and *Opportunity*, which landed on Mars in 2004. On the right is a test rover for the Mars Science Laboratory, which landed *Curiosity* on Mars in 2012.

Sojourner is 65 cm (2.13 ft) long. The Mars Exploration Rovers (MER) are 1.6 m (5.2 ft) long. *Curiosity* on the right is 3 m (9.8 ft) long.

Curiosity has an advanced payload of scientific equipment on Mars. It is the fourth NASA unmanned surface rover sent to Mars since 1996. Previous successful Mars rovers are *Sojourner* from the Mars Pathfinder mission (1997), and *Spirit* (2004–2010) and *Opportunity* (2004–present) rovers from the Mars Exploration Rover mission.

Curiosity is 2.9 m (9.5 ft) long by 2.7 m (8.9 ft) wide by 2.2 m (7.2 ft) in height, larger than Mars Exploration Rovers, which are 1.5 m (4.9 ft) long and have a mass of 174 kg (384 lb) including 6.8 kg (15 lb) of scientific instruments. In comparison to Pancam on the Mars Exploration Rovers, the MastCam-34 has 1.25× higher spatial resolution and the MastCam-100 has 3.67× higher spatial resolution.

The region the rover is set to explore has been compared to the Four Corners region of the North American west. Gale Crater has an area similar to Connecticut and Rhode Island combined.

Colin Pillinger, leader of the *Beagle 2* project, reacted emotionally to the large number of technicians monitoring *Curiosity*'s descent, because *Beagle 2* had only four people monitoring it. The *Beagle 2* team made a virtue out of necessity; it was known that there was no chance of obtaining funds in Europe, at that time, of the scale previously considered necessary for a Mars rover, so the team used innovative methods to reduce the cost to less than 4% of the cost of the *Curiosity* mission. They also had only one shot, with no funding for repeat missions (it was named *Beagle 2* as a successor to HMS *Beagle*, not to an earlier rover). It was considered a large risk, and, although *Beagle 2* did successfully survive its entry, descent, and landing, incomplete deployment of the solar panels hampered communication back to Earth. The team has proposed that a future launch might take multiple low-cost *Beagle*-type landers, with a realistic expectation that the vast majority would be successful, allowing exploration of several parts of Mars and possibly asteroids, all for considerably less cost than a single "normal" rover expedition.

Landing

Landing Site

Curiosity landed in Quad 51 (nicknamed Yellowknife) of Aeolis Palus in Gale Crater. The landing site coordinates are: 4°35′22″S 137°26′30″E4.5895°S 137.4417°E. The location has been named Bradbury Landing in honor of science fiction author Ray Bradbury. Gale crater, an estimated 3.5 to 3.8 billion-year-old impact crater, is hypothesized to have first been gradually filled in by sediments; first water-deposited, and then wind-deposited, possibly until it was completely covered. Wind erosion then scoured out the sediments, leaving an isolated 5.5-kilometer-high (3.4 mi) mountain, Aeolis Mons ("Mount Sharp"), at the center of the 154 km (96 mi) wide crater. Thus, it is believed that the rover may have the opportunity to study two billion years of Martian history in the sediments exposed in the mountain. Additionally, its landing site is near an alluvial fan, which is hypothesized to be the result of a flow of ground water, either before the deposition of the eroded sediments or else in relatively recent geologic history.

According to NASA, an estimated 20,000 to 40,000 heat-resistant bacterial spores were on *Curiosity* at launch, and as much as 1,000 times that number may not have been counted.

Curiosity and surrounding area as viewed by MRO/HiRISE. North is left. (August 14, 2012; enhanced colors)

Rover Role in the Landing System

NASA video describing the landing procedure. NASA dubbed the landing as "Seven Minutes of Terror".

Previous NASA Mars rovers became active only after the successful entry, descent and landing on the Martian surface. *Curiosity*, on the other hand, was active when it touched down on the surface of Mars, employing the rover suspension system for the final set-down.

Curiosity transformed from its stowed flight configuration to a landing configuration while the MSL spacecraft simultaneously lowered it beneath the spacecraft descent stage with a 20 m (66 ft) tether from the "sky crane" system to a soft landing—wheels down—on the surface of Mars. After the rover touched down it waited 2 seconds to confirm that it was on solid ground then fired several pyros (small explosive devices) activating cable cutters on the bridle to free itself from the spacecraft descent stage. The descent stage then flew away to a crash landing, and the rover prepared itself to begin the science portion of the mission.

Coverage, Cultural Impact and Legacy

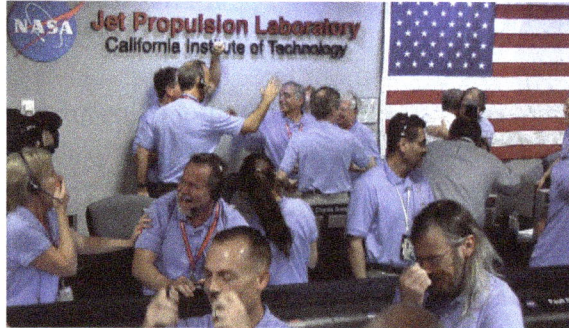

Celebration erupts at NASA with the rover's successful landing on Mars (August 6, 2012).

President Barack Obama congratulates NASA's *Curiosity* team (August 13, 2012).

Live video showing the first footage from the surface of Mars was available at NASA TV, during the late hours of August 6, 2012 PDT, including interviews with the mission team. The NASA website momentarily became unavailable from the overwhelming number of people visiting it, and a 13-minute NASA excerpt of the landings on its YouTube channel was halted an hour after the landing by a robotic DMCA takedown notice from Scripps Local News, which prevented access for several hours. Around 1,000 people gathered in New York City's Times Square, to watch NASA's live broadcast of *Curiosity*'s landing, as footage was being shown on the giant screen. Bobak Ferdowsi, Flight Director for the landing, became an Internet meme and attained Twitter celebrity status, with 45,000 new followers subscribing to his Twitter account, due to his Mohawk hairstyle with yellow stars that he wore during the televised broadcast.

On August 13, 2012, U.S. President Barack Obama, calling from aboard Air Force One to congratulate the *Curiosity* team, said, "You guys are examples of American know-how and ingenuity. It's really an amazing accomplishment." (Video (07:20))

U.S. flag on Mars

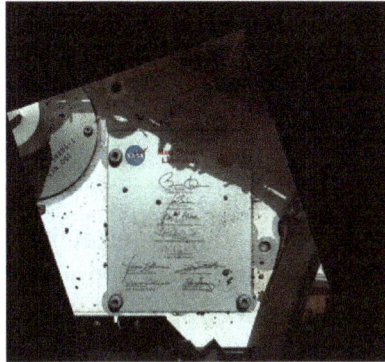

Plaque of President Obama and Vice President Joe Biden's signatures on Mars

Scientists at the Getty Conservation Institute in Los Angeles, California, viewed the CheMin instrument aboard *Curiosity* as a potentially valuable means to examine ancient works of art without damaging them. Until recently, only a few instruments were available to determine the composition without cutting out physical samples large enough to potentially damage the artifacts. CheMin directs a beam of X-rays at particles as small as 400 micrometres (0.016 in) and reads the radiation scattered back to determine the composition of the artifact in minutes. Engineers created a smaller, portable version named the *X-Duetto*. Fitting into a few briefcase-sized boxes, it can examine objects on site, while preserving their physical integrity. It is now being used by Getty scientists to analyze a large collection of museum antiques and the Roman ruins of Herculaneum, Italy.

Prior to the landing, NASA and Microsoft released *Mars Rover Landing*, a free downloadable game on Xbox Live that uses Kinect to capture body motions, which allows users to simulate the landing sequence.

NASA gave the general public the opportunity from 2009 until 2011 to submit their names to be sent to Mars. More than 1.2 million people from the international community participated, and their names were etched into silicon using an electron-beam machine used for fabricating micro devices at JPL, and this plaque is now installed on the deck of *Curiosity*. In keeping with a 40-year tradition, a plaque with the signatures of President Barack Obama and Vice President Joe Biden was also installed. Elsewhere on the rover is the autograph of Clara Ma, the 12-year-old girl from Kansas who gave *Curiosity* its name in an essay contest, writing in part that "curiosity is the passion that drives us through our everyday lives."

On August 6, 2013, *Curiosity* audibly played "Happy Birthday to You" in honor of the one Earth year mark of its Martian landing, the first time for a song to be played on another planet. This was also the first time music was transmitted between two planets.

On June 24, 2014, *Curiosity* completed a Martian year—687 Earth days—after finding that Mars once had environmental conditions favorable for microbial life. *Curiosity* will serve as the basis for the design of the Mars 2020 rover mission that is presently planned to be launched to Mars in 2020. Some spare parts from the build and ground test of *Curiosity* may be used in the new vehicle.

Awards

The NASA/JPL Mars Science Laboratory/*Curiosity* Project Team was awarded the 2012 Robert J. Collier Trophy by the National Aeronautic Association "In recognition of the extraordinary achievements of successfully landing *Curiosity* on Mars, advancing the nation's technological and engineering capabilities, and significantly improving humanity's understanding of ancient Martian habitable environments."

Curiosity's Location

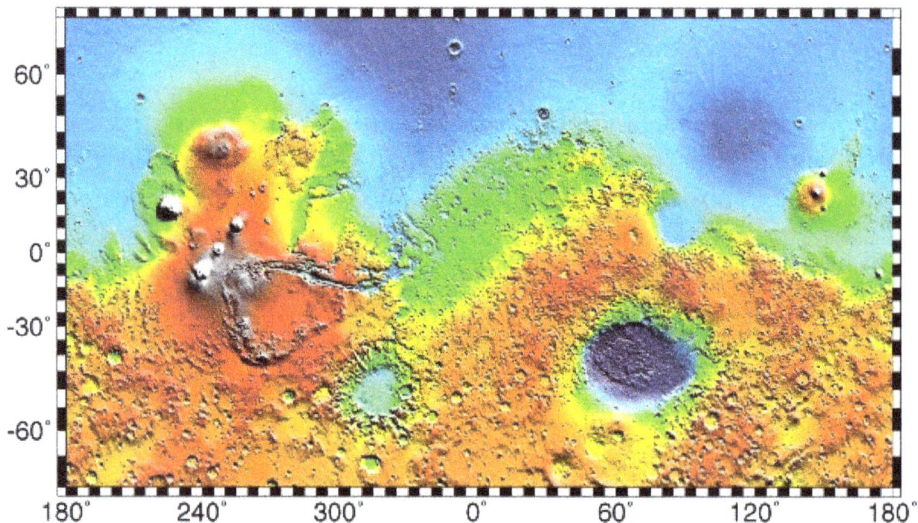

Interactive imagemap of the global topography of Mars, overlain with locations of Mars landers and rovers. Hover your mouse to see the names of prominent geographic features, and click to link to them. Coloring of the base map indicates relative elevations, based on data from the Mars Orbiter Laser Altimeter on NASA's Mars Global Surveyor. Reds and pinks are higher elevation (+3 km to +8 km); yellow is 0 km; greens and blues are lower elevation (down to −8 km). Whites (>+12 km) and browns (>+8 km) are the highest elevations. Axes are latitude and longitude; note poles are not shown.

Canadarm

The Shuttle Remote Manipulator System (SRMS), also known as Canadarm (Canadarm 1), is a series of robotic arms that were used on the Space Shuttle orbiters to deploy, maneuver and capture payloads. After the Space Shuttle Columbia disaster, the Canadarm was always paired with the Orbiter Boom Sensor System (OBSS), which was used to inspect the exterior of the Shuttle for damage to the thermal protection system.

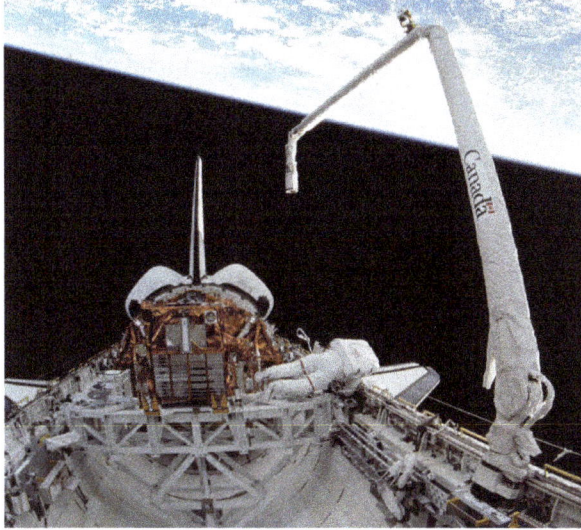

Canadarm (right) during Space Shuttle mission STS-72

Development

In 1969, Canada was invited by the National Aeronautics and Space Administration (NASA) to participate in the Space Shuttle program. At the time what that participation would entail had not yet been decided but a manipulator system was identified as an important component. Canadian company, DSMA Atcon, had developed a robot to load fuel into CANDU nuclear reactors; this robot attracted NASA's attention. In 1975, NASA and the Canadian National Research Council (NRC) signed a memorandum of understanding that Canada would develop and construct the Shuttle Remote Manipulator System.

NRC awarded the manipulator contract to Spar Aerospace. Three systems were constructed within this design, development, test and evaluation contract: an engineering model to assist in the design and testing of the Canadarm, a qualification model that was subjected to environmental testing to qualify the design for use in space, and a flight unit. Frank Mee is credited as the inventor of the Canadarm End Effector. Its design was inspired by the opening and closing of a camera's iris. His design won over the claw-like mechanisms that were being considered.

The main controls algorithms were developed by SPAR and by subcontractor Dynacon Inc. of Toronto. CAE Electronics Ltd. in Montreal provided the display and control panel and the hand controllers located in the Shuttle aft flight deck. Other electronic interfaces, servoamplifiers and power conditioners located on the Canadarm were designed and built by SPAR at its Montreal factory. The graphite composite boom that provides the structural connection between the shoulder and the elbow joint and the similar boom that connects the elbow to the wrist were produced by General Dynamics in the United States. Dilworth, Secord, Meagher and Associates, Ltd. in Toronto was contracted to produce the engineering model end effector then SPAR evolved the design and produced the qualification and flight units.The shuttle flight software that monitors and controls the Canadarm was developed in Houston, Texas, by the Federal Systems Division of IBM. Rockwell International's Space Transportation Systems Division designed, developed, tested and built the systems used to attach the Canadarm to the payload bay of the orbiter.

An acceptance ceremony for NASA was held at Spar's RMS Division in Toronto on the 11th of February 1981. Here Larkin Kerwin, then the head of the NRC, gave the SRMS the informal name, Canadarm.

The first remote manipulator system was delivered to NASA in April 1981. In all, five arms (arm 201, 202, 301, 302, and 303) were built and delivered to NASA. Arm 302 was lost in the Challenger accident.

Design and Capabilities

F. Story Musgrave, anchored on the end of the Canadarm, prepares to be elevated to the top of the Hubble Space Telescope during STS-61.

The original Canadarm was capable of deploying and retrieving payloads weighing up to 332.5 kg (733 lb) in space. In the mid-1990s the arm control system was redesigned to increase the payload capability to 3,293 kg (7,260 lb) in order to support space station assembly operations. While able to maneuver payloads with the mass of a loaded bus in space, the arm motors cannot lift the arm's own weight when on the ground. NASA therefore developed a model of the arm for use at its training facility within the Johnson Space Center located in Houston, Texas. The Canadarm can also retrieve, repair and deploy satellites, provide a mobile extension ladder for extravehicular activity crew members for work stations or foot restraints, and be used as an inspection aid to allow the flight crew members to view the orbiter's or payload's surfaces through a television camera on the Canadarm.

The basic Canadarm configuration consists of a manipulator arm, a Canadarm display and control panel, including rotational and translational hand controllers at the orbiter aft flight deck flight crew station, and a manipulator controller interface unit that interfaces with the orbiter computer. Most of the time, the arm operators see what they are doing by looking at the Advanced Space Vision System screen next to the controllers.

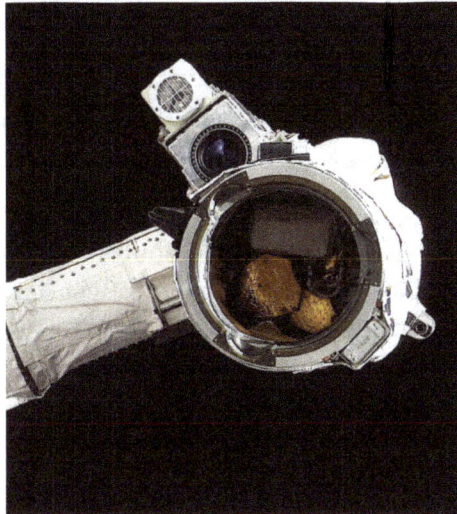

The Canadarm 1 End Effector

One crew member operates the Canadarm from the aft flight deck control station, and a second crew member usually assists with television camera operations. This allows the Canadarm operator to view Canadarm operations through the aft flight deck payload and overhead windows and through the closed-circuit television monitors at the aft flight deck station.

The Canadarm is outfitted with an explosive-based mechanism to allow the arm to be jettisoned. This safety system allows the Orbiter's payload bay doors to be closed in the event that the arm fails in an extended position and is not able to be retracted.

The Canadarm is 15.2 m (50 ft) long and 38 cm (15 in) diameter with six degrees of freedom. It weighs 410 kg (900 lb) by itself, and 450 kg (990 lb) as part of the total system. The Canadarm has six joints that correspond roughly to the joints of the human arm, with shoulder yaw and pitch joints, an elbow pitch joint, and wrist pitch, yaw, and roll joints. The end effector is the unit at the end of the wrist that grapples the payload's grapple fixture. The two lightweight boom segments are called the upper and lower arms. The upper boom connects the shoulder and elbow joints, and the lower boom connects the elbow and wrist joints.

Service History

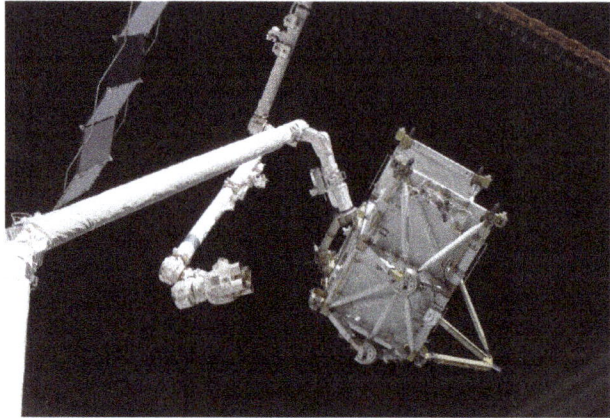

The Canadarm2 moves toward a P5 truss section, being held by Discovery's Canadarm, in preparation for a hand-off during STS-116

A simulated Canadarm installed on the Space Shuttle *Enterprise* was seen when the prototype orbiter's payload bay doors were open to test hangar facilities early in the shuttle program. The Canadarm was first tested in orbit in 1981, on Space Shuttle *Columbia*'s STS-2 mission. Its first operational use was on STS-3 to deploy and maneuver the Plasma Diagnostics Package. Canadarm has since flown on more than 90 missions with all five orbiters.

Since the installation of the Canadarm2 on the International Space Station, the two arms have been used to hand over segments of the station for assembly from the Canadarm to the Canadarm2; the use of both elements in tandem has earned the nickname of 'Canadian Handshake' in the media.

Retirement

The Canadarm's 90th and final shuttle mission was in July 2011 on STS-135, delivering the *Raffaello* MPLM to the ISS and back. *Discovery's* Canadarm is displayed next to her in the National Air and Space Museum. *Endeavour* left its OBSS at the International Space Station as part of its final mission, while its Canadarm was originally going to be displayed in the headquarters of the Canadian Space Agency. However, *Endeavour*'s Canadarm is now on permanent display at the Canada Aviation and Space Museum in Ottawa. The last of the Canadarms to fly in space, the SRMS flown aboard *Atlantis* on the final space shuttle mission, STS-135 in July 2011, was shipped to NASA's Johnson Space Center in Houston for engineering study and possible reuse on a future mission.

Da Vinci Surgical System

The da Vinci Surgical System (sic) is a robotic surgical system made by the American company Intuitive Surgical. Approved by the Food and Drug Administration (FDA) in 2000, it is designed to facilitate complex surgery using a minimally invasive approach, and is controlled by a surgeon from a console. The system is commonly used for prostatectomies, and increasingly for cardiac valve repair and gynecologic surgical procedures. According to the manufacturer, the da Vinci System is called "da Vinci" in part because Leonardo da Vinci's "study of human anatomy eventually led to the design of the first known robot in history."

da Vinci Surgical Systems operate in hospitals worldwide, with an estimated 200,000 surgeries conducted in 2012, most commonly for hysterectomies and prostate removals. As of June 30, 2014, there was an installed base of 3,102 units worldwide, up from 2,000 units at the same time the previous year. The location of these units are as follows: 2,153 in the United States, 499 in Europe, 183 in Japan, and 267 in the rest of the world. The "Si" version of the system costs on average slightly under US$2 million, in addition to several hundred thousand dollars of annual maintenance fees. The da Vinci system has been criticised for its cost and for a number of issues with its surgical performance.

Overview

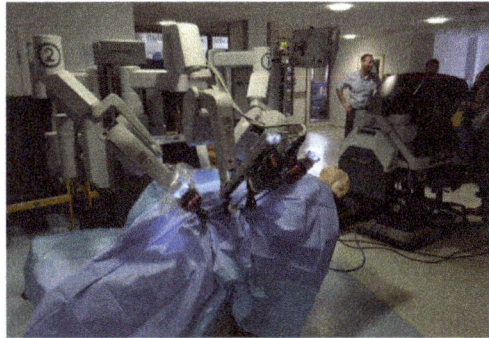

da Vinci patient-side component (left) and surgeon console (right)

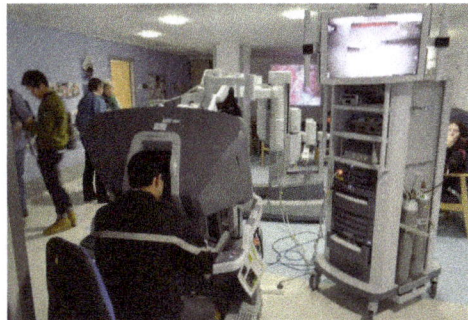

A surgeon console at the treatment centre of Addenbrooke's Hospital

The da Vinci System consists of a surgeon's console that is typically in the same room as the patient, and a patient-side cart with four interactive robotic arms controlled from the console. Three of the arms are for tools that hold objects, and can also act as scalpels, scissors, bovies, or unipolar or hi. The surgeon uses the console's master controls to maneuver the patient-side cart's three or four robotic arms (depending on the model). The instruments' jointed-wrist design exceeds the natural range of motion of the human hand; motion scaling and tremor reduction further interpret and refine the surgeon's hand movements. The da Vinci System always requires a human operator, and incorporates multiple redundant safety features designed to minimize opportunities for human error when compared with traditional approaches.

The da Vinci System has been designed to improve upon conventional laparoscopy, in which the surgeon operates while standing, using hand-held, long-shafted instruments, which have no wrists. With conventional laparoscopy, the surgeon must look up and away from the instruments, to a nearby 2D video monitor to see an image of the target anatomy. The surgeon must also rely on a patient-side assistant to position the camera correctly. In contrast, the da Vinci System's design allows the surgeon to operate from a seated position at the console, with eyes and hands positioned in line with the instruments and using controls at the console to move the instruments and camera.

By providing surgeons with superior visualization, enhanced dexterity, greater precision and ergonomic comfort, the da Vinci Surgical System makes it possible for more surgeons to perform minimally invasive procedures involving complex dissection or reconstruction. For the patient, a da Vinci procedure can offer all the potential benefits of a minimally invasive procedure, including less pain, less blood loss and less need for blood transfusions. Moreover, the da Vinci System can enable a shorter hospital stay, a quicker recovery and faster return to normal daily activities.

FDA Clearance

The Food and Drug Administration (FDA) cleared the da Vinci Surgical System in 2000 for adult and pediatric use in urologic surgical procedures, general laparoscopic surgical procedures, gynecologic laparoscopic surgical procedures, general non-cardiovascular thoracoscopic surgical procedures and thoracoscopically assisted cardiotomy procedures. The FDA also cleared the da Vinci System to be employed with adjunctive mediastinotomy to perform coronary anastomosis during cardiac revascularization.

Representative clinical uses

The da Vinci System has been successfully used in the following procedures:

- Radical prostatectomy, pyeloplasty, cystectomy, nephrectomy and ureteral re-implantation;

- Hysterectomy, myomectomy and sacrocolpopexy;

- Hiatal hernia repair;

- Spleen-sparing distal pancreatectomy, cholecystectomy, Nissen fundoplication, Heller myotomy, gastric bypass, donor nephrectomy, adrenalectomy, splenectomy and bowel resection;

- Internal mammary artery mobilization and cardiac tissue ablation;

- Mitral valve repair and endoscopic atrial septal defect closure;

- Mammary to left anterior descending coronary artery anastomosis for cardiac revascularization with adjunctive mediastinotomy;

- Transoral resection of tumors of the upper aerodigestive tract (tonsil, tongue base, larynx) and transaxillary thyroidectomy

- Resection of spindle cell tumors originating in the lung

Future Applications

Although the general term "robotic surgery" is often used to refer to the technology, this term can give the impression that the da Vinci System is performing the surgery autonomously. In contrast, the current da Vinci Surgical System cannot – in any manner – function on its own, as it was not designed as an autonomous system and lacks decision making software. Instead, it relies on a human operator for all input; however, all operations – including vision and motor functions— are performed through remote human-computer interaction, and thus with the appropriate weak AI software, the system could in principle perform partially or completely autonomously. The difficulty with creating an autonomous system of this kind is not trivial; a major obstacle is that surgery *per se* is not an engineered process – a requirement for weak AI. The current system is designed merely to replicate seamlessly the movement of the surgeon's hands with the tips of micro-instruments, not to make decisions or move without the surgeon's direct input.

The possibility of long-distance operations depends on the patient having access to a da Vinci System, but technically the system could allow a doctor to perform telesurgery on a patient in another country. In 2001, Dr. Marescaux and a team from IRCAD used a combination of high-speed fiber-optic connection with an average delay of 155 ms with advanced asynchronous transfer mode (ATM) and a Zeus telemanipulator to successfully perform the first transatlantic surgical procedure, covering the distance between New York and Strasbourg. The event was considered a milestone of global telesurgery, and was dubbed "Operation Lindbergh".

Criticism

Critics of robotic surgery assert that it is difficult for users to learn and that it has not been shown to be more effective than traditional laparoscopic surgery. The da Vinci system uses proprietary software, which cannot be modified by physicians, thereby limiting the freedom to modify the operation system. Furthermore, its $2 million cost places it beyond the reach of many institutions.

The manufacturer of the system, Intuitive Surgical, has been criticized for short-cutting FDA approval by a process known as "premarket notification," which claims the product is similar to already-approved products. Intuitive has also been accused of providing inadequate training, and encouraging health care providers to reduce the number of supervised procedures required before a doctor is allowed to use the system without supervision. There have also been claims of patient injuries caused by stray electrical currents released from inappropriate parts of the surgical tips used by the system. Intuitive counters that the same type of stray currents can occur in non-robotic laparoscopic procedures. A study published in the *Journal of the American Medical Association* found that side effects and blood loss in robotically-performed hysterectomies are no better than those performed by traditional surgery, despite the significantly greater cost of the system. As of 2013, the FDA is investigating problems with the da Vinci robot, including deaths during surgeries that used the device; a number of related lawsuits are also underway.

From a social analysis, a disadvantage is the potential for this technology to dissolve the creative freedoms of the surgeon, once hailed by scholar Timothy Lenoir as one of the most professional individual autonomous occupations to exist. Lenoir claims that in the "heroic age of medicine," the surgeon was hailed as a hero for his intuitive knowledge of human anatomy and his well-crafted techniques in repairing vital body systems. Lenoir argues that the da Vinci's 3D console and robotic arms create a mediating form of action called medialization, in which internal knowledge of images and routes within the body become external knowledge mapped into simplistic computer coding.

Sensei Robotic Catheter System

The Sensei X robotic catheter is a medical robot designed to enhance a physician's ability to perform complex operations using a small flexible tube called a catheter. As open surgical procedures that require large incisions have given way to minimally invasive surgeries in which the surgeon gains access to the target organs through small incisions using specialized surgical tools. One important tool used in many of these procedures is a catheter used to deliver many of things a surgeon needs to do his work, to impact target tissue and deliver a variety of medicines or disinfecting agents to treat disease or infection.

Manufactured by Hansen Medical, Sensei X is a specialized robotic catheter system that is controlled by a physician and is designed for accurate positioning, manipulation and stable control of catheter and catheter-based technologies during cardiovascular procedures. The Sensei system obtained U.S. Food and Drug Administration (FDA) clearance in 2007, after which the Cleveland Clinic's electrophysiology program, then directed by Dr. Andrea Natale, received the first placement. The Sensei Robotic Catheter System and Artisan Extend Control Catheter allow physicians to navigate flexible catheters with greater stability and control during complex cardiac arrhythmia procedures. Hansen Medical has related co-development agreements with the following industry leaders: St. Jude Medical, GE Healthcare, Siemens Healthcare, and Philips Medical Systems.

Hansen Medical was founded by Dr. Frederic Moll, who had also co-founded Intuitive Surgical, manufacturer of the Da Vinci Surgical System, whose use has propelled medicinal robotics to the forefront of patient care. The Sensei system is indicated for use during the cardiac mapping phase of cardiac arrhythmia treatment in the US. Meanwhile, it has CE mark approval for facilitating the navigation of ablation catheters within the atria of the heart during complex arrhythmia procedures such as Atrial Fibrillation (AF).

In November 2010, Hansen Medical received unconditional Investigational Device Exemption (IDE) approval from the FDA initiating a clinical trial to investigate the use of the Sensei X Robotic Catheter System and the Artisan Control Catheter for treatment of AF, the most common cardiac arrhythmia. The Principal Investigator of the ARTISAN AF Trial is Dr. Andrea Natale, executive medical director for Texas Cardiac Arrhythmia Institute (TCAI). The first case in the trial was completed by Dr. Joseph Gallinghouse, an electrophysiologist, at the TCAI at St. David's Medical Center.

By the end of 2010, nearly 100 Sensei systems have been shipped worldwide since its 2007 release. The system has been used to perform almost 5,000 procedures. The Sensei system operates by guiding standard catheters through a manipulated robotically steerable sheath (hollow tube) in the patient's vasculature. The doctor performs the procedure at a control station with a technology called "IntelliSense" to proximally measure the forces applied along the shaft of the catheter as a result of catheter tissue contact. The Artisan catheter has two robotically controlled segments which provides six degrees of freedom and 270 degrees of bend articulation which can assist physicians in accessing hard-to-reach cardiac anatomy. The open lumen Artisan catheter accommodates 8F percutaneous EP catheters. Centers have reported acute and long term success rates consistent with manual procedures.

The Sensei Robotic System in Clinical use

Although the Sensei system was initially tested in a range of ablation procedures including SVT and typical atrial flutter, there is most excitement about its role in complex ablation procedures such as for atrial fibrillation (AF), where the ability to manipulate catheters to precise locations within the heart and keep them stable in the desired position is crucial. Achieving adequate tissue contact, ideally with a small amount of pressure being applied by the catheter during ablation, is also essential to effectively destroy the heart tissue responsible for arrhythmia. The ability to titrate catheter contact force using the Sensei system's built-in pressure sensor technology (called intellisense) may allow operators to maximise the chance of creating effective burns across the thickness of the atrial wall, whilst minimising the risk of complication. Intracardiac echocardiography has also demonstrated that greater catheter stability is achieved with robotic compared to manual catheter delivery. Consequently, there is evidence that robotic ablation causes more effective and more efficient burns.

Use of robotic navigation for catheter ablation was also designed to allow Electrophysiology to perform most of the procedure without being exposed to X-rays (or radiation). The radiation dose to the operator during a conventional manual ablation procedure is relatively small, although there is cumulative exposure that becomes an important consideration for operators performing procedures on a daily basis. By performing procedures a few metres away using a robotic system and seated in the leaded anteroom, the operator is shielded from X-rays and is less vulnerable to operator fatigue, which may affect operator performance in long complex cases. Use of robotic navigation has been shown to reduce fluoroscopy times in catheter ablation of AF, resulting in reduced X-ray exposure for patients and other health care professionals present in the catheter laboratory.

Early Experience with the Sensei Robotic System

The system has been the subject of several clinical trials in the USA and Europe, particularly for the catheter ablation of AF. There were early safety concerns after the 'first-in-human' studies suggested high complication rates. Wazni et al. reported experience with the first 71 catheter ablations for AF in their centre using the Hansen robotic navigation system starting from 2005.

These early studies have allowed others to incorporate changes to their technique, and hence recent work has produced complication rates for catheter ablation of AF comparable to procedures performed manually. The field of robotic ablation is growing and evolving rapidly, and randomised controlled trials comparing robotic to manual ablation are ongoing in Europe and the USA to see if these potential advantages will translate into better clinical outcomes.

Recent Studies Using the Sensei Robotic System

Since there are no completed randomised controlled trials comparing robotic ablation to manual ablation, comparing the modalities remains difficult. Techniques, complication rates and clinical results vary widely between centres for both manual and robotic procedures, making registry data difficult to compare. Typical ranges for procedures performed manually are: major complication rates of 3-5%, and success rates of approximately 80-90% for paroxysmal AF and 70-75% for persistent AF (depending on the length of follow-up). The following are recent reports in medical journals detailing experience of the Sensei system in the catheter ablation of AF, and demonstrate approximate success and complication rates at the time of writing:

Study	Parameter	Robotic ablation	Manual ablation
Prague 2011	Number of subjects	100 total All paroxysmal	(No comparator group)
	Pulmonary vein isolation achieved acutely	100%	
	Procedure time	3.7 ± 0.9 hours	
	Fluoroscopy time	11.9 ± 7.8 minutes	
	Clinical success (Freedom from AF)	63% at 15 ± (3-28) month, 86% after 21 patients had a repeat procedure	
	Safety	Total complications 0%	
Texas Group 2009	Number of subjects	193 total 135 paroxysmal 55 persistent 6 long-lasting persistent	197 total (registry cohort) 127 paroxysmal 55 persistent 11 long-lasting persistent
	Pulmonary vein isolation achieved acutely	100%	100%
	Procedure time	3.1 ± 0.8 hours	3.1 ± 0.8 hours
	Fluoroscopy time	48.9 ± 24.6 minutes	58.4 ± 20.4 minutes
	Clinical success (Freedom from AF)	Overall 85% at 14 ± 1 month Paroxysmal AF 90% Persistent AF 71% Long-lasting 100% (of only 6)	Overall 81% at 14 ± 1 month Paroxysmal AF 85% Persistent AF 73% Long lasting 67%
	Safety	Total complications 1.5% 1% tamponade	Total complications 1.0% 0.5% tamponade
Hamburg 2010	Number of subjects	64 total (all paroxysmal)	(No comparator group)
	Pulmonary vein isolation achieved acutely	100%	
	Procedure time	3.0 (2.5- 3.8) 1.5 hours	
	Fluoroscopy time	24 (12-34) minutes	
	Clinical success (Freedom from AF)	81% at 12 months	
	Safety	Total complications 0%	

Hamburg 2009	Number of subjects	65 total Paroxysmal 43 Persistent 22	
	Pulmonary vein isolation achieved acutely	95% (remainder completed manually)	
	Procedure time	3.3 ± 0.7 hours	
	Fluoroscopy time	17 ± 7 minutes	
	Clinical success (Freedom from AF)	73% at 8 months 76% Paroxysmal 78% persistent	
	Safety	Total complications 5% Tamponade 1.5%	
Prague 2009	Number of subjects	22 total All paroxysmal	16 total (registry cohort) All paroxysmal
	Pulmonary vein isolation achieved acutely	100%	100%
	Procedure time	3.5 ± 0.5 hours	4.2 ± 1.0 hours
	Fluoroscopy time	15 ± 5 minutes	27 ± 9 minutes
	Clinical success (Freedom from AF)	91% at 5 ± 1 month	81% at 9 ± 3 month
	Safety	Total complications 0%	Total complications 0%
Multicenter group (France, Germany, Italy, Prague, USA) 2008	Number of subjects	40 total Paroxysmal 29 Persistent 11	(No compartor group)
	Pulmonary vein isolation achieved acutely	100%	
	Procedure time	2.8 ± 1.5 hours	
	Fluoroscopy time	64 ± 33 minutes	
	Clinical success (Freedom from AF)	98% at 12 months	
	Safety	Total complications 5% Tamponade 5% (nil else)	

References

- Mann, Adam (August 7, 2012). "The Photo-Geek's Guide to Curiosity Rover's 17 Cameras". Wired. Retrieved January 16, 2015.

- Klinger, Dave (August 7, 2012). "Curiosity says good morning from Mars (and has busy days ahead)". Ars Technica. Retrieved January 16, 2015.

- Vieru, Tudor (December 6, 2013). "Curiosity's Laser Reaches 100,000 Firings on Mars". Softpedia. Retrieved January 16, 2015.

- Moskowitz, Clara (January 7, 2013). "NASA's Curiosity Rover Brushes Mars Rock Clean, a First". Space.com. Retrieved January 16, 2015.

- Clark, Stuart (January 17, 2015). "Beagle 2 spacecraft found intact on surface of Mars after 11 years". The Guardian. Retrieved January 18, 2015.

- Chang, Kenneth (October 5, 2015). "Mars Is Pretty Clean. Her Job at NASA Is to Keep It That Way.". The New York Times. Retrieved October 6, 2015.

- Hoover, Rachel (June 24, 2014). "Ames Instrument Helps Identify the First Habitable Environment on Mars, Wins Invention Award". NASA. Retrieved June 25, 2014.

- Canadian Space Agency (2 May 2013). "Minister Moore Unveils Exhibit for Canada's National Space Icon: the Canadarm". Canadian Space Agency. Retrieved 1 July 2013.

- Clark, Stephen (November 17, 2011). "Nuclear power generator hooked up to Mars rover". Spaceflight Now. Retrieved November 11, 2013

- Brown, Dwayne; Cole, Steve; Webster, Guy; Agle, D.C. (August 22, 2012). "NASA Mars Rover Begins Driving at Bradbury Landing". NASA. Retrieved August 22, 2012.

- Martin, Paul K. (June 8, 2011). "NASA'S MANAGEMENT OF THE MARS SCIENCE LABORATORY PROJECT (IG-11-019)" (PDF). NASA Office of Inspector General. Retrieved August 6, 2012.

- Amos, Jonathan (August 3, 2012). "Gale Crater: Geological 'sweet shop' awaits Mars rover". BBC News. Retrieved August 6, 2012.

- Amos, Jonathan (August 17, 2012). "Nasa's Curiosity rover prepares to zap Martian rocks". BBC News. Retrieved September 3, 2012.

- "Rover Environmental Monitoring Station for MSL mission" (PDF). 4th International workshop on the Mars Atmosphere: modelling and observations. Pierre und Marie Curie University. February 2011. Retrieved August 6, 2012.

Permissions

Index

www.ingramcontent.com/pod-product-compliance
Lightning Source LLC
Chambersburg PA
CBHW061935190326
41458CB00009B/2742